The Theory of the Earth
By Baron Cuvier

PREFACE

Geology, now deservedly one of the most, popular and attractive of the physical sciences, was, not many years ago, held in little estimation and even at present, there are not wanting some who do not hesitate to maintain, that it is a mere tissue of ill observed phenomena, and of hypotheses of boundless extravagance. The work of Cuvier now laid before the public, contains in itself not only a complete answer to these ignorant imputations, but also demonstrates the accuracy, extent, and importance of many of the facts and reasonings of this delightful branch of Natural History. Can it be maintained of a science, which requires for its successful prosecution an intimate acquaintance with Chemistry, Natural Philosophy and Astronomy,—with the details and views of Zoology. Botany, and Mineralogy, and which connects these different departments of knowledge in a most interesting and striking manner,—that it is of no value? Can it be maintained of Geology, which discloses to us the history of the first origin of organic beings, and traces their gradual development from the monade to man himself,—which enumerates and describes the changes that plants, animals, and minerals—the atmosphere, and the waters of the globe—have undergone from the earliest geological periods up to our own time, and which even instructs us in the earliest history of the human species,—that it offers no gratification to the philosopher? Can even those who estimate the value of science, not by intellectual desires, but by practical advantages, deny the importance of Geology, certainly one of the foundations of agriculture, and which enables us to search out materials for numberless important economical purposes?

Geology took its rise m the Academy of Freyberg, with the illustrious Werner, to whom we owe its present interesting condition This being the case, we ought not, (as is at present too much the practice), amidst the numerous discoveries in the mineral kingdom which have been made since the system of investigation of that great interpreter of nature was made known, forget the master, and arrogate all to ourselves. In this Island, Geology first took firm root in the north : in Edinburgh the Wernerian geognostical views and method of investigation, combined with the theory of Hutton, the experiments and speculations of Hall, the illustrations of Playfair, and the labours of the Royal and Wernerian Natural History Societies, excited a spirit of inquiry which rapidly spread throughout the Empire; and now Great Britain presents to the scientific world a scene of geological acuteness, activity, and enterprise, not surpassed in any other country.

On the Continent the writings of Cuvier, distinguished equally by purity and beauty of style and profound learning, have proved eminently useful in aiding the progress of Geology. In this country Cuvier was first made known as a geologist by the publication of the present essay, which, from its unexampled popularity, has made his name as familiar to us as that of the most distinguished of our own writers.

ROBERT JAMESON,
College Museum, Edinburgh, 25th November 1826.

PRELIMINARY OBSERVATIONS.

In my work on Fossil Bones, the object which I proposed was to discover to what animals the osseous remains, with which the superficial strata of the globe are ffiled, may have belonged. In pursuing this object, I had to fallow a path in which but little progress had hitherto been made. As an antiquary of a new order, I was obliged at once to learn the art of restoring these monuments of past revolutions to their original forms, and to discover theb nature and relations; I had to collect and bring together in their original order, the fiagments of which they consisted; to reconstruct, as it were, the ancient beings to which these fragments belonged; to reproduce them with all their proportions and characters; and, lastly, to compare them with those which now live at the surface of the globe :—an art almost unknown, and which presupposed a science whose first developments had scarcely yet been traced, that of the laws which regulate the coexistence of the forms of the different parts in organised beings. I had therefore to prepare myself for these inquiries, by others of a far more extensive kind, respecting the animals which still exist.

Nothing, except an almost complete review of creation n its present state, could give a character of demonstration to the results of my investigation into its ancient state; but, from this review, I had at the same time to expect a great body of rules and affinities not less satisfactorily demonstrated; and it became obvious, that, in consequence of this essay upon a small portion of the theory of the earth, the whole animal kingdom would necessarily be in some measure subjected to new laws.

Thus I was encouraged in this twofold investigation by the equal interest which it promised to possess, both with regard to the general science of anatomy, the essential basis of all those which treat of organised bodies, and with regard to the physical history of the globe, the foundation of mineralogy, geography, and even, it may be saidy of the history of Man, and of all that it most concerns him to know with regard to himself.

If it be so interesting to us to follow, in the infancy of our species, the almost obliterated traces of extinct nations, why should it not also be so, to search, amid the darkness of the infancy of the Earth, for the traces of revolutions which have taken place anterior to the existence of all nations? We admire the power by which the human mind has measured the motions of the celestial bodies, which nature seemed to have concealed for ever from our view. Geuius and science have burst the limits of space; and observations, explained by just reasoning, have unveiled the mechanism of the universe, Would it not also be glorious for man to burst the limits of time, and, by means of observations, to ascertain the history of this world, and the succession of events which preceded the birth of the human race? Astronomers have undoubtedly advanced more rapidly than naturalists; and the present period, with respect to the Theory of the Earthr bears some resemblance to that which some philosophers fancied that, the heavens were formed of polished stones, and that the moon was of the size of the Peloponnesus; but after Anaxagoras, came Copernicus and Kepeer, who pointed the way to Newton; and why should not natural history also one day have its Newton?

Plan of this Essay

What I especially propose to present in this discourse, is the plan and the result of my labours regarding Fossil Bones. I shall also attempt to trace a rapid sketch of the efforts that have been made up to the present day, to restore the history of the revolutions of the globe. The facts which I have been enabled to discover, form, without doubt, only a small portion of those which would be necessary to complete this ancient history: but several of them lead to decisive consequences and the vigorous manner in which I have proceeded in their determination, affords me reason to think that they will be regarded as points definitively fixed, and which in their aggregate will form an epoch in science. Lastly, I trust their novelty will be a sufficient excuse for me. if I claim for them the earnest attention of my readers.

My object will first be to show by what relations the history of the fossil bones of terrestrial animals connects itself with the theory of the earth, and for what reasons a peculiar importance is to be attributed to it, with reference to this subject. I shall then unfold the principles upon which is founded the art of determining these bones, or, in other words, of recognising a genus, and of distinguishing a species, by a single fragment of bone,—an art, on the certainty of which depends that of my whole work. I shall give a rapid account of the new species, and of genera previously unknown, which the application of these principles has led me to discover, as well as the different kinds of deposits in which they are contained. And as the difference between these species and those which exist at the present day is bounded by certain limits, I shall show that these limits much exceed those which now distinguish the varieties of the same species. I shall therefore make known to what extent these varieties may go, whether from the influence of time, or from that of climate, or, lastly, from that of domestication.

In this way I shall be enabled to conclude, and to induce my readers to conclude with me, that great events were necessary to produce the more considerable differences which I have discovered. I shall next mention the particular modifications which my researches must necessarily introduce into the hitherto received opinions regarding the revolutions of the globe; and, lastly, I shall inquire how far the civil and religious history of different nations corresponds, with the results of observation with regard to the physical history of the Earth, and with the probabilities which these observations afford concerning the period at which societies of men may have found fixed places of abode, and fields susceptible of cultivation, and at which, therefore, they may have assumed a durable form.

First Appearance of the Earth,

When the traveller passes over those fertile plains where gently flowing streams nourish in their course an abundant vegetation, and where the soil, inhabited by a numerous population, adorned with flourishing villages, opulent cities, and superb monuments, is never disturbed, except by the ravages of war, or by the oppression of the powerful, he is not led to suspect that Nature also has had her intestine wars, and that the surface of the globe has been broken up by revolutions and catastrophes. But his ideas change as soon as he digs into that soil which now presents so peaceful an aspect, or ascends to the hills which border the plain; his ideas are expanded, if I may use the expression, in proportion to the expansion of the view, and begin to embrace the full extent and grandeur of those ancient events, when he climbs the more elevated chains whose base is skirted by these hills, or when, by following tho beds of the torrents which descend from those chains, he penetrates, as it were, into their interior.

The lowest and most level parts of the earth, exhibit nothing, even when penetrated to a very great depth, but horizontal strata composed of substances more or less varied, and containing almost all of them innumerable marine productions. Similar strata, with the same kind of productions, compose the lesser hills to a considerable height. Sometimes the shells are so numerous as to constitute of themselves the entire mass of the rock; they rise to elevations superior to the level of every part of the ocean, and are found in placcs where, no sea could have carried them at the present day, under any circumstances; they are not only enveloped in loose sand, but are often inclosed in the hardest rocks. Every part of the earth, every hemisphere, every continent, every island of any extent, exhibits the same phenomenon.

The times are past when ignorance could maintain, that these remains of organised bodies are mere sportings of nature, productions generated ia the womb of the Earth, by its own creative powers; and the efforts made by some metaphysicians of the present day, will not probably succeed in bringing these exploded opinions again into repute. A scrupulous comparison of the forms of these remains, of their texture, and often even of their chemical composition, does not disclose the slightest difference between the fossil shells and those Which still inhabit the sea: the preservation of the former is not less perfect than that of the latter; most commonly we neither observe detrition nor fracture in them, nothing, in short, that announces a violent removal from their original places; the smallest of them retain their sharpest ridges, and their most delicate spines. They have, therefore, not only lived in the sea, but they have also been deposited by it. It is the sea which has left them in the places where they are now found. But this sea has remained for a certain period in those places; it has covered them long enough, and with sufficient tranquillity to form thosee deposits, so regular, so thick, so extensive, and partly also so solid, which contain those remains of aquatic animals. The basin of the sea has therefore undergone one change at least, either in extent, or in situation. Such is the result of the very first search, and of the most superficial examination.

The traces of revolutions become still more apparent and decisive, when we ascend a little higher, and approach nearer to the foot of the great chains. There are still found many beds of shells; some of these are even thicker and more solid; the shells are quite as numerous, and as well preserved, but they are no longer of the same species. The strata which contain them are not so generally horizontal; they assume an oblique position, and are sometimes almost vertical. While in the plain and low hills it was necessary to dig deep in order to discover the succession of the beds, we here discover it at once by their exposed edges, as we follow the valleys that have been I produced by their disjunction. Great masses of debris form at the foot of the cliffs, rounded hills, the height of which is augmented by every thaw and tempest.

These inclined strata, which form the ridges of the secondary mountains, do not rest upon the horizontal strata of the hills which are situate at their base, and which form the first steps in approaching them; but, on the contrary, dip under them, while the hill in question rest upon their declivities. When we dig through the horizontal strata in the vicinity of mountains whose strata are inclined, we find these inclined strata reappearhig below; and even sometimes, when the inclined strata are not too elevated, their summit is crowned by horizontal ones. The inclined strata are therefore older than the horizontal strata; and as they must necessarily, at least the greater number of them, have been formed in a horizontal position, it is evident that they have been raised, and that this change in their direction has been effected before the others were superimposed upon them.

Thus the sea, previous to the deposition of the horizontal strata, had formed others, which, by the operation of problematical causes, were broken, raised, and overturned in a thousand ways; and, as several of those inclined strata which it had formed at more remote periods, rise higher than the horizontal strata which have succeeded them, and which surround them, the causes by which the inclination of these beds was effected, had also made them project above the level of the sea, and formed islands of them, or at least shoals and inequalities; and this must have happened, whether they had been raised by one extremity, or whether the depression of the opposite extremity had made the waters subside. This is the second result, not less clear, nor less satisfactorily demonstrated, than the first, to every one who will take the trouble of examining the monuments on which it is established.

Proofs that such revolutions have been numerous.

But it is not to this subversion of the ancient strata, nor to this retreat of the sea after the formation of the new strata, that the revolutions and changes which have given rise to the present state of the Earth are limited.

When we institute a more detailed comparison between the various strata and those remains of animals which they contain, we presently perceive, that this ancient sea has not always deposited mineral substances of the same kind, nor remains of animals of the same species; and that each of its deposits has not extended over the whole surface which it covered. There has existed a succession of variations; the former of which alone have been more or less general, while the others appear to have been much less so. The elder the strata are, the more uniform in each of them over a great extent; the newer they are, the more limited are they, and the more subject to vary at small distances. Thus the displacements of the strata were accompanied and followed by changes in the nature of the fluid, and of the matters which it held in solution; and when certain strata, by making their appearance above the waters, had divided the surface of the seas by islands and projecting ridges, different changes might take plate in particular basins.

Amidst these variations in the nature of the general fluid, it is evident, that the animals which lived in it could not remain the same. Their species, and even their genera, changed with the strata; and, although the same species occasionally recur at small distances, it may be announced as a general truth, that the shells of the ancient strata have forms peculiar to themselves; that they gradually disappear, so as no longer to be seen at all in the recent strata, and stil less in the presently existing ocean, in which their corresponding species are never discovered, and where several, even of their genera, do not occur : that, on the contrary, the shells of the recent strata are similar, in respect to their genera, to those which exist in our seas; and that, in the latest and least consolidated of these strata and in certain recent and limited deposits, there are some species which the most experienced eye could not distinguish from those which are found in the neighbouring seas.

There has, therefore, been a succession of variations in the economy of organic nature, which has been occasioned by these of the fluids in which the animals lived, or which has at least corresponded with them; and these variations have gradually conducted the classes of aquatic animals to their present state, till, at length, at the time when the sea retired from our continents for the last time, its inhabitants did not differ much from those which are found in it at the present day.

We say for the last time, because, if we examine with still greater care those remains of organised bodies, we discover, in the midst of even the oldest strata of marine formation, other strata replete with animal or vegetable remains of terrestrial or fresh-water productions; and, amongst the more recent strata, or in other words, those that are nearest the surface, there are some in which land animals are buried under heaps of marine productions. Thus, the various catastrophes which have disturbed the strata, have not only caused the different parts of our continents to rise by degrees from the bosom of the waves, and diminished the extent of the basin of the ocean, but have also given rise to numerous shiftings of this basin. It has frequently happened, that lands which have been laid dry, have been again covered by the waters. In consequence either of their being ingulphed in the abyss, or of the sea having merely risen over them. The particular portions also, of the Earth, which the sea abandoned in its last retreat, those which are now inhabited by man and terrestrial animals,—had already been once laid dry and had then afforded subsistence to quadrupeds, birds, plants, and land productions of all kinds: the sea which left it had, therefore, covered it at a previous period.

The changes in the level of the waters have not, therefore, consisted solely in a more or less gradual, or more or less general retreat; there have been various successive irruptions and retreats, the linal result of which, however, has been a universal depression of the level of the sea Proofs that these Revolutions have been sudden.

It is of much importance to remark, that these repeated irruptions and retreats of the sea have neither all been slow nor gradual; on the contrary; most of the catastrophes which have occasioned them have been sudden; and this is especially easy to be proved, with regard to the last of these catastrophes, that which, by a two-fold motion, has inundated, and afterwards laid dry, our present continents, or at least a part of the land which forms them at the present day. In the northern regions, it has left the carcases of large quadrupeds which became enveloped in the ice, and have thus been preserved even to our own times, with their skin, their hair, and their flesh. If they had not been frozen as soon as killed, they would have been decomposed by putrefaction. And, on the other hand, this eternal frost could not previously have occupied the places in which they have been seized by it, for they could not have lived in such a temperature. It was, therefore, at one and the same moment that these animals were destroyed, and the country which they inhabited became covered with ice. This event has been sudden, instantaneous, without any gradation; and what is so clearly demonstrated with respect to this last catastrophe, is not less so with reference to those which have preceded it. The breaking to pieces, the raising up and overturning of the older strata, leave no doubt upon the mind that they have been reduced to the state in which we now see them, by the action of sudden and violent causes; and even the force of the motions excited in the mass of waters, is still attested by the heaps of debris and rounded pebbles which are in many places interposed between the solid strata. Life, therefore, has often been disturbed on this earth by terrible events. Numberless living beings have been the victims of these catastrophes; some, which inhabited the dry land, have been if wallowed up by inundations; others, which peopled the waters, have been laid dry, from the bottom of the sea having been suddenly raised; their very races have been extinguished for ever, and have left no other memorial of their existence than some fragments, which the naturalist can scarcely recognize.

Such are the conclusions to which we are necessarily led by the objects that we meet with at every step, and which we can always verify, by examples drawn from almost every country. These great and terrible events are every where distinctly recorded, so as to be always legible by the eye skilled to decypher their history in the monuments which they have left behind.

But what is still more astonishing and not less certain, life has not always existed upon the globe; and it is easy for the observer to distinguish the point at which it has begun to deposit its production.

Proofs that there have been Revolutions anterior to the existence of living beings.

If we ascend to higher points of elevation, and advance towards the great ridges, the craggy summits of the mountain chains, we shall presently find those remains of marine animals, those innumerable shells, of which we have spoken, becoming more rare, and at length disappearing alto-gether. We arrive at strata of a different nature, which contain no vestiges of living beings. Nevertheless, their crystallization, and even their stratification, shew that they have been also in a liquid state at their formation ; their inclined position, and the cliffs into which they are broken, show that they also have been forcibly moved from their original places; the oblique manner in which they dip ander the shelly strata, that they have been formed previously to these latter; and lastly, the height to which their rugged and bare peaks rise above all these shelly strata, that their summits had already emerged from the waters, when the shelly strata were forming.

Such are those celebrated Primitive Mountains which traverse our continents in different directions, raising themselves above the clouds, separating the basins of rivers from one another, affording, in their perennial snows, reservoirs which feed the springs, and forming, in some measure, the skeleton, and as it were the rough framework, of the Earth.

The eye perceives from afar, in the indentations with which their ridge has been marked, and in the sharp peaks with which it is bristled, indications of the violent manner in which they have been elevated. Their appearance, in this respect, is very different from that of those rounded mountains, and hills with long flat surfaces, whose less ancient masses have always remained in the situation in which they were quintly deposited by the waters of more recent seas.

These indications become more obvious as we approach. The valleys have n0 longer those gently-sloping sides, those salient and re-entering angles corresponding on either side to each other, which seem to denote the beds of ancient streams. They widen and they contract without any general rule; their waters, at one time, expand into lakes: at another, fall in torrents; and sometimes their rocks, suddenly approaching from each side, form transverse dikes, over which the waters tumble in cataracts. The dissevered strata, while they show on one side their edges perpendicularly raised, on the other present large portions of their surface lying obliquely; they do not correspond in height, but those which, on one side, form the summit of the cliff, often dip underneath on the other, and are no longer visible.

Yet, amidst all this confusion, distinguished naturalists have been able to demonstrate, that there still reigns a certain order, and that those immense deposits, broken and overturned though they be, observe a regular succession with regard to each other, which is nearly the same in all the great mountain chains. According to them, Granite, of which the central ridges of the greater number of these chains consist, and which thus surmounts every other rock, is also the rock which is found deepest in the solid crust of the globe. It is the most ancient of those which we have found means of examining in the place assigned them by nature; and we inquire not at present, whether it owes its origin to a general fluid, which formerly held every thing in solution, or may have been the first consolidated by the cooling of a great mass in fusion, or even in a state of vapour. Foliated rocks rest upon its sides, and form the lateral ridges of these great chains; schists, porphyries, sandstones, and talcose rocks, intermingle with their strata; lastly, granular marbles, and other limestones destitute of shells, resting upon the schists, form the outer ridges, the lower steps as il were, the counterforts, of these chains, and are the last formations, by which this unknown fluid, this sea without inhabitants, would seem to have prepared materials for the mollusca and zoophytes, which were presently to deposite upon these foundations vast heaps of their shells and corals.

We even find the first productions of these mollusca and zoophytes appearing in small numbers. and scattered at greater or less distances, in the last strata of these primitive formations, or in that portion of the crust of the globe to which geologists have given the name of Transition rocks. Here and there we meet with beds containing shells, interposed between certain granites of later formation than the others, between schists of various kinds, and between some newer beds of granular marbles. Life, which was in the end to obtain entire possession of the globe, seems, in these primordial times, to have struggled with the nert nature which formerly predominated; and it was not until a considerable time after, that it obtained the ascendancy over it, and acquired for itself the exclusive right of continuing and elevating the solid envelope of the Earth.

Hence, it is impossible to deny, that the masses which now constitute our highest mountains, have been originally in a liquid state; and that they have for a long time been covered by waters in which no living beings existed. Thus, it has not been only since the appearance of life that changes have been operated in the nature of the matters which have been deposited; for the masses formed previous to that event, have varied, as well as those which have been formed since. They have also experienced violent changes in their position, and a part of these changes must have taken place at the period when these masses existed by themselves, and were not covered over by the shelly masses. The proof of this lies in the overturnings, the disruptions, and the fissures which are observable in their strata, as well as in those of more recent formations, and which are in the ancient strata even in greater number and better defined.

But these primitive masses have also undergone other revolutions since the formation of the secondary strata, and have, perhaps, given rise to, or at least have partaken of, some of those changes which these strata themselves have experienced. There are actually considerable portions of the primitive formations uncovered, although placed in lower situations than many of the secondary formations; and we cannot conceive how it should have so happened, unless the primitive strata in those places had forced themselves into view, after the secondary strata had been formed. In certain countries, we find numerous large blocks of primitive substances scattered over the surface of secondary formations, and separated by deep valleys, or even by arms of the sea, from the peaks or ridges from which they must have been derived. We must necessarily conclude, therefore, either that these blocks have been ejected by eruptions, or that the valleys (which must have stopped their course) did not exist at the time of their being transported; or, lastly, that the motions of the waters by which they were transported, exceeded in violence any thing that we can imagine at the present day.

Here, therefore, we have a collection of facts, a series of epochs, anterior to the present time, of which the successive steps may be perfectly ascertained, although the duration of their intervals cannot be defined with precision. They are so many fixed points, which serve to regulate and direct our inquiries respecting this ancient chronology.

2.Examination of the Causes which act at present on the surface of the Globe.

Let us now examine those changes which are taking place at the present day upon the globe, investigating the causes which still act in its surface, and endeavouring to determine the possible extent of their effects. This portion of the history of the Earth is so much the more important, that it has long been considered possible to explain the more ancient revolutions on its surface by means of these still existing causes; a same manner as it is found easy to explain past events in political history, by an acquaintance with the passions and intrigues of the present day. But we shall presently see, that unfortunately the case is different in physical history —the thread of operations is here broken; the march of Nature is changed; and none of the agents which she now employs, would have been sufficient for the production of her ancient works.

There still exist, however, four causes in full activity, which contribute to alter the surface of our continents. These are, rains and thaws, which waste down the steep mountains, and precipitate the fragments to their bottoms running waters, which carry off these fragments, and deposit them in places where their current is abated; the sea, which undermines the foundations of elevated coasts, forcing steep cliffs, and which throws up great banks of sand upon the low coasts; and, lastly, volcanoes, which pierce through the solid strata from below, elevate these strata, or spread over the surface vast quantities of ejected matter

In every place where the broken strata present their edges on abrupt surfaces, there fall down to their base, every spring, and even after every storm, fragments of their materials, which are rounded by rolling upon each other. These collected heaps gradually assume an inclination determined by the laws of cohesion, and thus form, at the bottom of the cliff, taluses, of greater or less elevation, according as the fragments which have fallen are more or less abundant. These taluses constitute the sides of the valleys in all elevated, mountainous regions, and are covered with a rich vegetation, whenever the fragments from the upper parts begin to fall less abundantly; but their want of solidity subjects themselves also to slips, when they are undermined by rivulets. On these occasions, towns, and rich and populous districts, are sometimes buried under the ruins of a mountain; the courses of rivers are interrupted, and lakes are formed in places which were before the abodes of fertility and cheerfulness. Fortunately these great slips happen but seldom, and the principal use of those hills of debris, is to furnish materials for the ravages of torrents.

Alluvial Formations

The rains which fall, the vapours which are condensed, and the snows which are melted, upon the ridges and summits of mountains, descend, by an infinite number of rills, along their slopes, carrying with them some portions of the materials of which these slopes are composed, and tracing slight furrows by their passage. These rills soon unite in the deeper gutters with which the surface is marked, run off by the deep valleys which intersect their bottom, and thus form streams and rivers, which carry back to the sea the waters it had formerly supplied to the atmosphere. On the melting of the snows, or when a storm takes place, these mountain torrents become suddenly swollen, and rush down the declivities with a velocity proportioned to their steepness. They dash violently against the bases of those taluses of fallen fragments which cover the sides of all the high valleys, earning off the already rounded fragments of which they are composed, and which thus become smoothed, and still farther polished, by attrition. But in proportion as they reach the more level valleys, where their violence is diminished, or when they arrive at more expanded basins, where their waters are permitted to spread, they throw out upon their banks the largest of those stones which they had rolled down. The smaller fragments arc deposited still lower; and nothing reaches the great canal of the river excepting the minutest particles, or the most impalpable mud. It often happens, also, that before these streams unite to form great rivers, they have to pass through large and deep lakes, in which their mud is deposited, and from which tln ir waters come forth limpid.

The lower rivers, and all the streams which descend from the less elevated mountains and hills, also produce effects, upon the districts through which they flew, more or less analogous to those of the torrents from the higher mountains. When these rivers are swollen by great rains, they attack the base of the earthy or sandy hills which they meet with in their course, and carry their fragments to be deposited upon the lower grounds, and which are thus, in some degree, raised by each succeeding inundation. Finally, when the rivers reach great lakes or the sea, and when that rapidity, which carried off and kept in suspension the particles of mud comes to cease entirely, these particles are deposited at the sides of their mouths, where they form low grounds, by which the shores are prolonged. And if these shores are such, that the sea also throws up sand upon them, and thus contributes to their increase; there are created, as it were, provinces, and even entire kingdoms, which usually become the most fertile, and speedily the richest, in the world, if their rulers permit human industry to exert itself in peace

Formation of Downs.

The effects which the sea produces, without the cooperation of rivers, are much less beneficial. When the coast is low, and the bottom sandy, the waves push the sand toward the shore, where, at every reflux of the tide, it becomes partially dried; and the wind, which almost always blows from the sea, drifts it upon the beach. Thus are formed those hillocks of sand, named Downs, which, if the industry of man does not fix them by suitable plants, move slowly, but invariably, toward the interior of the country, and overwhelm fields and dwellings, because the same wind that raises the sand of the beach upon the down, throws that of its summit in the opposite direction from the sea. When the nature of the sand, and that of the water which is raised with it, are such as to form a durable cement, the shells and bones, thrown upon the beach, become incrusted with life. Pieces of wood-trunks of trees, and plants growing near the sea, are enveloped in these aggregates; and thus are produced what might be denominated indurated downs, such as we see upon the coasts of New Holland, and of which a precise idea may be formed from the description given of them by Peron.

Formation of Cliffs or Steep Shores.

On the other hand, when the coast is high, the sea, which is thus prevented from throwing up any thing, exercises a destructive action upon it. Its waves, by sapping the foundation, cause the superincumbent portion of the face of the cliff, thus deprived of support, to be incessantly falling down in fragments. These fragments are tumbled about by the billows, until the softer and more divided parts disappear. The harder portions, from being rolled in contrary directions, assume the form of boulders and pebbles; and these, at length, accumulate in sufficient quantity to form a rampart, by which the bottom of the cliff is protected against farther depredations

Such is the action of water upon the solid land; and we see, that it consists almost entirely in reducing it to lower levels, but not indefinitely.

The fragments of the great mountain ridges are carried down into the valleys; their finer particles, together with those of the lower hills and plains, are borne to the sea; alluvial depositions extend the coasts at the expence of the high grounds. These are limited effects, to which vegetation in general puts a stop, and which, besides, presuppose the existence of mountains, valleys, and plains, in short, all the inequalities of the globe; and which, therefore, cannot have given rise to these inequalities. The formation of downs is a phenomenon still more limited, both ift regard to height and horizontal extent: and has no relation whatever to that of those enormous masses into the origin of which it is the object of geology to inquire.

Depositions formed in Water.

Although we cannot obtain a precise knowledge of the action exerted by water within its own bosom, it is yet possible to determine its limits to a certain degree

Lakes, pools, marshes, and sea-ports, into which rivulets discharge their waters, more especially when these descend from near and steep hills, deposit large quantities of mud, which would at length fill them up entirely, if care were not taken to clean them out. The sea also throws quantities of slime and sediment into harbours and creeks; into all places, in short, where its waters are more tranquil than ordinary. The currents also heap up at their meeting, or throw out at their sides, the sand which they are continually raising from the bottom of the sea, forming it into banks and shallows.

Stalactites.

Certain waters, after dissolving calcareous substances by means of the superabundant carbonic acid with which they are impregnated, allow these substances to crystallize after the acid has evaporated; and in this manner form stalactites and other concretions. There are strata, confusedly crystallized in fresh water, which are sufficiently extensive to be compared with some of those which have been deposited by the ancient sea. The famous Travertine quarries of the neighbourhood of Rome, and the rocks of the same substance, which are formed, and continually varied in figure, by the river of Teverona, are generally known. These two modes of action may be combined; the deposits accumulated by the sea may be solidified by stalactite. Thus, when springs abounding in calcareous matter, or containing some other substance in solution, happen to fall into places where these deposits are formed, we then find aggregates in which marine and fresh-water productions may be blended. Of this description are the banks in the island of Guadeloupe, which, along with human skeletons, present land and sea shells mingled together. Of the same nature also is the quarry described by Saussure, in the neighbourhood of Messina, in which the sand-stone is seen forming by the consolidation of the sand thrown up by the sea.

Lithophytes.

In the torrid zone, where lithophytes of many species abound, and are propagated with great rapidity, their strong trunks are interwoven and accumulated so as to form rocks and reefs; and rising even to the surface of the water, shut up the entrance of harbours, and lay frightful snares for navigators. The sea, throwing up sand and mud upon the tops of these shoals, sometimes raises their surface above its own level, and forms islands, which are soon covered with a rich vegetation

Incrustation.

It is also possible, that, in particular places, large quantities of the animals inhabiting shells, leave their stony coverings when they die, and that these, cemented together by slime of greater or less consistence, or by other cementing substances, form extensive deposits or shell banks. But we have no evidence that the sea can now incrust those shells with a paste as compact as that of the marbles, the sandstones, or even the coarse limestone (calcaire grossier) in which we see the shells of our strata enveloped. Still less do we any where find the sea depositing those more solid and more siliceous strata which have preceded the formation of the shelly strata.

In short, all these causes united, would not change, in an appreciable degree, the level of the se ; nor raise a single stratum above its surface; and still less would they produce the smallest hillock upon the surface of the earth.

It has been asserted that the sea has undergone a general diminution of level; and proofs of this are said to have been discovered in some parts of the shores of the Baltic. But whatever may be the causes of these appearances, we are certain that they are not general in their operation; and that, in the greater number of harbours, where any alteration of the level would be a matter of so much interest, and where fixed and ancient works afford so many means of measuring its variations, the mean level of the sea is constant. There has, therefore, never been a universal lowering, nor a universal encroachment, of the waters of the ocean. In some places, indeed, such as Scotland, and various parts of the Mediterranean, evidence has been thought to have been found, that the sea has risen, and that it now covers shores which were formerly above its level.

Volcanoes.

The action of volcanoes is still more limited, and more local, than any of these which have yet been mentioned. Although we have no precise idea of the means by which nature keeps up these violent fires at such great depths, we can judge decidedly, by their effects, of the changes which they may have produced at the surface of the globe. After a volcano has announced itself, by some shocks of an earthquake, it forms for itself an opening. Stones and ashes are thrown to a great distance, and lava is vomited forth. The more fluid part of the lava flows in long streams, while the less fluid portion stops at the edges of the opening, raises its margins all round, and forms a cone, terminated by a crater. Thus volcanoes accumulate upon the surface matters which were previously buried in the bowels of the earth, after modifying their nature, and raise themselves into mountains. By these means, they have formerly covered some parts of our continent, and have also suddenly produced islands in the middle of the sea. But these mountains and islands have always been composed of lava, and all their materials have undergone the actiion of fire: they are disposed as matters should be, which have flowed from an elevated point. Volcanoes, therefore, neither raise nor overturn the strata through which their apertures pass: and if some causes acting from those depths have contributed, in certain cases, to raise up large mountains, they cannot have been volcanic agents of the same nature as those which exist at the present day.

Thus, we repeat, it is in vain that we search, among the powers which now act at the surface of the earth, for causes sufficient to produce the revolutions and catastrophes, the traces of which are exhibited by its crust. And if we have recourse to the constant external forces with which we are as yet acquainted, we shall have no greater success.

Constant Astronomical Causes.

The pole of the earth moves in a circle around the pole of the ecliptic, and its axis is more or less inclined to the plane of the ecliptic; but these two motions, the causes of which are now ascertained, are much too limited for the productien of effects like those whose magnitude we have just been stating. At any rate, their excessive slowness would render them altogether inadequate to account for catastrophes which, as we have shown, must have been sudden.

The same reasoning applies to all other slow-motions which have been conceived as causes of the revolutions in question, chosen doubtless in the hope that their existence could not be denied, because it might always be easy to hold out that their very slowness rendered them imperceptible. But whether they be true or not is of little importance, for they explain nothing, as no cause acting slowly could have produced sudden effects.

Admitting that there has been a gradual diminution of the waters; that the sea has transported solid matters in all directions; that the temperature of the globe is either diminishing or increasing; none of these causes could have overturned our strata; enveloped in ice large animals, with their flesh and skin; laid dry marine testacea, the shells erf which arc, at the present day, as well preserved as if they had been drawn up alive from the sea; and, lastly, destroyed numerous species, and even entire genera.

These considerations have struck most naturalists; and among those who have endeavoured to explain the present state of the globe, hardly any one has attributed it entirely to the agency of slow causes, still less to causes operating under our eyes. The necessity to which they are thus reduced, ofseeking for causes different from those which we see acting at the present day, is the very circumstance that has forced them to make so many extraordinary suppositions, and to lose themselves in so many erroneous and contradictory speculations, that the very name of their science, as I have elsewhere remarked, has long been a subject of ridicule to prejudiced persons, who have only looked to the systems which It has been the means of hatching, and have forgotten the extensive and important series of authentic facts which it has brought to light.

3.Older Systems of Geologists.

During a long time, two events or epochs only, the Creation and the Deluge, were admitted as comprehending the changes which have been operated upon the globe; and all the efforts of geologists were directed to account for the present existing state of things, by imagining a certain original state, afterwards modified by the deluge, of which also, as to its causes, its operations, and its effects, each entertained his own theory.

Thus, according to one, the earth was at first invested with an uniform light crust, which covered the abyss of the sea; and which being broken up for the production of the deluge, formed the mountains by its fragments. According to another, the deluge was occasioned by a momentar suspension of cohesion among the particles of mineral bodies; the whole mass of the globe was dissolved, and the paste thus formed became penetrated with shells. According to a third, God raised up the mountains for the purpose of allowing the waters, which had produced the deluge, to run off: and selected those places in which there was the greatest quantity of rocks, without which the mountains could not have supported themselves. A fourth created the earth from the atmosphere of one comet, and deluged it by the tail of another: The heat which it retained from its origin, was what, in his opinion, excited the whole of the living beings upon it to sin; for which they were all drowned, excepting the fishes, whose passions were apparently less vehement.

It is evident, that, even while confined within the limits prescribed by the Book of Genesis, naturalists might still have a pretty wide range : they soon found themselves, however, in too narrow bounds; and when they had succeeded in converting the six days of creation into so many indefinite periods, the lapse of ages no longer forming an obstacle to their views, their systems took a flight proportioned to the periods which they could then dispose of at pleasure.

Even the great Leibnitz amused himself, like Descartes, by conceiving the earth to be an extinguished sun, a vitrified globe, upon which the vapours falling down again, after it had cooled, formed seas, which afterwards deposited the lime-stone formations.

By Demaillet the whole globe was conceived to have been covered with water for many thousands of years. He supposed this water had gradually retired; that all the land animals were originally inhabitants of the sea; that man himself commenced his career as a fish; and he asserts, that is not uncommon, even now, to meet with fishes in the occan, which are still only half converted into men, but whose descendants will in time become perfect human beings.

The system of Buffon is merely an extension of that of Leibnitz, with the addition only of a comet, which, by a violent blow, struck off from the sun the liquefied mass of the earth, together with those of all the other planets at the same instant. From this supposition, he was enabled to assume positive dates, as, from the present temperature of the earth, it could be calculated how long it had taken to cool down so far; and, as all the other planets had come from the sun at the same time, it could also be calculated how many ages are still required for cooling the greater ones, and to what degree the smaller are already frozen.

More recent Systems.

In our own times, men of still bolder imaginations have exercised their minds upon this great subject. Some writers have revived and greatly extended the ideas of Demaillet. They suppose that every thing was originally fluid; that this fluid gave existence to animals, which were at first of the most simple kind, such as the monads and other infusory and microscopic species; that, in process of time, and by assuming different habits, the races of animals became complicated, and assumed that diversity of nature and character in which they now appear. By means of those various races of animals, part of the waters of the sea have gradually been converted into calcareous earth; while the vegetables, concerning the origin and metamorphoses of which these writers are totally silent, have, on their part, converted a portion of the same water into clay. These two earths, on being shipped of the characters which life had impressed upon them, are resolved, by a final analysis, into silex; and hence the reason that the oldest mountains are more siliceous than the rest. All the solid parts of the earth, therefore, owe their existence to life, and, without life; the globe would still be entirely liquid.

Other writers have preferred the ideas of Kepler, and, like that great astronomer, have considered the globe itself as possessed of vital faculties. According to them a vital fluid circulates in it; a process of assimilation goes on in it, as well as in animated bodies; every particle of it is alive; it possesses instinct and volition, even to the most elementary molecules, which attract and repel each other according to sympathies and antipathies. Each kind of mineral has the power of converting immense masses into its own nature, as we convert our food into flesh and blood. The mountains are the respirators organs of the globe, and the schists its organs of secretion; it is by these latter that it decomposes the water of the sea, in order to produce the matters ejected by volcanoes. The veins are carious sores, abscesses of the mineral kingdom; and the metals are products of rottenness and disease, which is the reason that almost all of them have so bad a smell.

More recently still, a philosophy, which substitutes metaphor for reasoning, and proceeds on the system of absolute identity or of pantheism, attributes the production of all phenomena, or which, in the eyes of its supporters, is the same thing, all beings, to polarization, such as is manifested by the two electricities; and denominating every kind of opposition or difference, whether of situation, of nature, or of function, by the title of Polarisation, opposes to each other, in the first place, God and the universe; then, in the universe, the sun and the planets; next, in each planet, the solid and the liquid; and, pursuing this course, changing its figures and allegories according to its necessities, at length arrives at the last details of organic species.

It must, however, be observed, that these are what may be termed extreme examples, and that all geologists have not carried the extravagance of their conceptions to such a length as those which we have just cited. Yet, among those who have proceeded with more caution, and have not searched for geological causes beyond the limits of physical and chemical science, much diversity and contradiction still prevail

Diversities of all the Systems.

According to one system, every thing has been successively precipitated by crystallization, and deposited nearly as it exists at present; but the sea, which covered all has gradually retired.

According to another, the materials of which the mountains consist, are incessantly worn down and carried off by the rivers to be deposited at the bottom of the sea, where they are heated under an enormous pressure, and form strata, which are one day to be viololently lifted up by the heat which consolidates them.

A third supposes the fluid divided into a multitude of lakes, placed, like the seats of an amphitheatre, above each other, which, after having deposited our shelly strata, have successively broken their dikes, to descend and fill the basin of the ocean.

According to a fourth, tides of seven or eight hundred fathoms depth have carried off, from time to time, the matter lying at the bottom of the sea, and have thrown it in the form of mountains and hills, upon the original valleys or plains of the continent

A fifth makes the various fragments of which the earth is composed, fall successively from heaven, in the manner of meteoric stones, bearing the impress of their foreign origin in the unknown beings whose remains they contain.

A sixth represents the globe as hollow, and places within it a loadstone nucleus, which a transported from one pole to the other, by the attraction of comets, carrying along with it the centre of gravity, and the mass of waters at the surface; thus alternately drowning the two hemispheres.

We might mention twenty other systems, as different from one another as those enumerated. And to prevent mistake, we may here state, that our intention is not captiously to criticize or find fault with their authors; on the contrary, we admit that these ideas have generally been conceived by men of intellect and knowledge, who were not ignorant of facts, several of whom had even travelled extensively for the purpose of examining them, and who, ill this manner, made numerous and mportant additions to science.

Cause of these differences.

Whence comes it, then, that there should be so much contrariety in the solutions of the same problem, that are given by men who proceed upon the same principles? May not this hare been occasioned by the conditions of the problem never having been all taken into consideration at once; by which it has remained hitherto indeterminate, and susceptible of many solutions,—all equally good, when such or such conditions are abstracted; and all equally bad, when a new condition comes to be known, or when the attention is directed to some condition which had been formerly neglected?

To quit the language of mathematics, it may be asserted, that almost all the authors of these systems, confining their attention to certain difficulties which struck them more forcibly than others, have endeavoured to solve these in a manner more or less plausible, and have left unnoticed others, equally numerous, and equally important. For example, the only difficulty with one consisted in explaining the changes that had taken place in the level of the sea; with another, it consisted in accounting for the solution of all terrestrial substances in one and the same menstruum: and with a third, in showing how animals that were believed to be natives of the torrid zone could live in the frigid zone. Exhausting all the powers of the mind upon these questions, they conceived that they had done every thing that was neces-sary when they had contrived some method of answering them; and yet, while they neglected all the other phenomena, they did not always think of determining with precision the measure and limits of those which they had endeavoured to explain.

This is peculiarly the case with regard to the secondary formations, which constitute, however, the most important and most difficult part of the problem. During a long time, all that was done with respect to these, consisted of feeble attempts to determine the order of superposition of their strata, and the connections of these strata with the species of animals and plants whose remains they contain.

Are there certain animals and plants peculiar to certain strata, and not found iu others? What are the species that appear first in order, and what those which succeed? Do these two kinds of species sometimes accompany each other? Are there alternations in their appearance; or, in other words, do the first reappear a second time, and do the others then disappear? Have these animals and plants all lived in the places where their remains are found, or have they been trasported thither from other places? Do they all live at the present day in some part of the earth, or have they been partially or totally destroyed? Is there any constant connection between the antiquity of the strata and the resemblance, or non-resemblance, of the fossils contained in thsin to the animals and plants which now exist? Is there any connexion, in regard to climate, between the fosoils and such living beings as resemble them most? May it be concluded, that the transportation of these living beings, if such a thing ever happened, has taken place from north to south, or from east to west; or were they irregularly scattered and mingled together; and can the epochs of these transportations be determined by the characters which they have impressed upon the strata?

What can be said regarding the causes of the existing state of the globe, if no reply can be made to these questions,— if there be no sufficient grounds to determine the choice between answering in the affirmative or negative? It is but too true, that, for a long time, none of these points was satisfactorily determined; and scarcely even would geologists seem to have had any idea of the propriety of clearing them up before constructing their systems.

The reason of this strange procedure will be discovered, when we reflect, that ill geologists have hitherto been either mere cabinet naturalists, who had themselves paid little atttcntion to the structure of mountains, or mere mineralogists, who had not studied in sufficient detail the innumerable varieties of animals, and the infinite complication of their various parts. The former of these have only constructed systems: the latter have furnished excellent observations, and have laid the foundation of true geological science; but have been unable to complete the edifice.

4.Progress of Mineral Geology

The purely mineral part of the great problem of the Theory of the Earth has been investigated with admirable care by Saussure, and has been since earned to an astonishing degree of development by Werner, and by the numerous enlightened pupils of his school

The former of these celebrated men, by a laborious investigation of the most accessible districts, continued for twenty years, in which he examined the Alps on all sides, and penetrated through all their defiles; has laid open to our view the entire disorder of the primitive formations, and has distinctly traced the limits by which they are distinguished from the secondary formations. The other, taking advantage of the numerous excavations made in the most ancient mining district in the world, has fixed the laws by which the succession of the strata are regulated, pointing out the relative antiquity of these strata, and tracing each of them through all its metamorphoses. It is from him, and from him alone, that we date the commencement of real geology, in so far as concerns the mineral nature of the strata : but neither he nor Saussure have determined the fossil organic species occurring in each kind of stratum, with the accuracy which has become necessary, now that the number of animals already known is so great.

Other naturalists, it is true, have examined the fossil remains of organised bodies; they have collected and figured them by thousands, and their works will serve as so many precious collections of materials. But, considering these animals and plants more with reference to their own nature, than as connected with the theory of the earth; or regarding these petrifactions as curiosities, rather than as historical documents; or, lastly, contenting themselves with practical explanations regarding the position of each fragment, they have almost always neglected to investigate the general laws affecting the geological position of organic remains, or their connection with the strata.

Importance of Fossil Remains in Geology,

And yet the idea of such an investigation was very natural; for it is abundantly obvious, that it is to these fossil remains alone that we owe even the commencement of a theory of the earth, and that, without them, we should perhaps never have even suspected that there had existed any successive epochs, and a series of different operations, in the formation of the globe, by them alone we are, in fact, enabled to ascertain, that the globe has not always had the same external crust; because, we are thoroughly assured, that the plants and animals must have lived at the surface before they had thus come to be buried deep beneath it. It is only by analogy that we have been enabled to extend to the primitive formations, the conclusion which is furnished directly for the secondary by the organic remains which they contain; and if there had only existed formations in which no fossil remains were inclosed, it could never have been shewn that these formations had not all been of simultaneous origin.

It is also by means of the organic remains, slight as is the knowledge we have hitherto acquired of them, that we have been enabled to discover the little that we yet know respecting the nature of the revolutions of the globe. From them we have learned, that the strata in which they are buried have been quietly deposited in a fluid; that their variations have corresponded with those of the fluid in question; that their being laid bare has been occasioned by the transportation of this fluid to some other place; and that this circumstance must have befallen them more than once. Nothing of all this could have been known with certainty, had no fossil remains existed.

The study of the mineral part of geology, though not less necessary, and even of much more utility to the practical arts, is yet much less instructive with reference to the object of our present inquiry.

We remain in utter ignorance respecting the causes vthich have given rise to the variety in the mineral substances of which the strata are composed. We are even ignorant of the agents which may have held some of these substances in solution; and it is still disputed, respecting several of them, whether they have owed their origin to water or to fire. After all philosophers are only agreed on one point, which is, that the sea has changed its place; and how should this have been known, unless by means of the fossil remains?

The organic remains, therefore, which have given rise to the theory of the earth, have, at the same time, furnished it with its principal illustrations :—the only ones, indeed, that have as yet been generally acknowledged.

It is this consideration which has encouraged us to nvestigate the subject. But the field is vast; and it is but a very small portion of it that could be cultivated by the labour of a single individual. It was necessary, therefore, to select a particular department; and the choice was soon made. The class of fossil remains which forms the subject of this work, engaged our attention at the very outset, because it appeared to us to be that which is the most fertile in precise results, and yet, at the same lime, less known, andricher in new objects of research

High importance of the Fossil Bones of Quadrupeds,

It is obvious, in fact, that the fossil bones of quadrupeds must lead to more accurate conclusions than any other remains of organized bodies, and that for several reasons.

In the first place, they indicate much more clearly the nature of the revolutions to which they have already been subjected. Shells certainly announce the fact, that the sea has once existed in the places where they have been formed; but the changes which have taken place in their species, when vigorously inquired into, may have arisen from slight changes in the nature of the fluid in which they lived, or merely in its temperature. They may even have, been produced by causes still more accidental. We can never be perfectly assured that certain species, and even genera inhabiting the bottom of the sea, and occupying certain fixed spaces, for a longer or shorter time, may not have been ditven away and supplanted byr other species or genera.

In regard to quadrupeds, on the contrary, every thing is precise. The appearance of their bones in strata, and still more of their entire carcases, announces, either that the stratum itself which contains them has, at a former period, been laid dry, or, at least, that dryland must have existed in its neighbourhood. Their disappearance renders it certain, that this stratum has been inundated, or that the dry land in question has ceased to exist. It is from them, therefore, that we learn with perfect certainty the important fact of repeated eruptions of the sea, which the shells and other marine productions could not of themselves have proved; and it is by a careful investigation of them, that we may hope to ascertain the number and the epochs of these irruptions

Secondly, the nature of the revolutions which have altered the surface of the globe, must have exerted a more powerful action upon terrestrial quadrupeds, than upon manne animals. As these revolutions have consisted chiefly of changes in the bed of the sea, and as the waters must have destroyed all the quadrupeds which they reached, if their irruption was general, it would necessarily have destroyed the entire class; or if it only overwhelmed certain continents at one time, it would at least have destroyed the species peculiar to those continents, without having the same effect upon the marine animals. On the other hand, millions of aquatic animals would have been left dry, or buried under newly-formed strata, or thrown violently on the coasts; while their races would still have been preserved in some more peaceful parts of the sea, whence they might again be propagated after the agitation of the waters had ceased.

Thirdly, this more complete action is also more easily ascertained. It is more easy to demonstrate its effects, because, the number of quadrupeds being limited, and the greater part of their species, at least the large ones, being known, we have more means of determining whether fossil bones belong to them, or to a species that is now lost. As, on the other hand, we are very far from being acquainted with all the testaceous animals and fishes which inhabit the sea, and as we are still probably ignorant of the greater number of these which live in deep water, it is impossible to know with certainty, whether a species which occurs in a fossil state, may not still exist somewhere live. And hence, we see naturalists persisting in giving the name of pelagic shells, that is to say, shells inhabiting the open sea, to the belemnites, cornua-ammonis, and other testaceous remains which have hitherto been found only in the older strata; meaning by this, that if they have not yet been discovered in a living state, it is because they inhabit the depths of the sea, far beyond the reach of our nets.

Small probability of discovering New Species of large Quadrupeds.

Naturalists, certainly, have not yet explored all the continents, nor do they even knov all the quadrupeds which inhabit the countries that they have explored. New species of this class are discovered from time to time and those who have not examined with attention all the circumstances belonging to these discoveries might also imagine that the unknown cuadrupeds, whose bones are found in our strata, may remain to this day concealed, in some islands not yet discovered by navigators, or in some of the vast deserts which occupy the middle of Asia, Africa, the two Americas, and New Holland.

However, if we carefully examine what kinds of quadrupeds have been recently discovered, and in what circumstances they have been found, we shall see that there is little hope of our ever finding alive those which have hitherto been observed only a a fossil state.

Islands of moderate extent, and at a considerable distance from the continents or large islands, possess very few quadrupeds, and these, for the greater part, of diminutive size. When they happen to contain any of the larger species, these must have been carried to them from other countries. Bougainville and Cook found no other large quadrupeds than hogs and dogs in the South Sea Islands; and the largest species of the West India Islands was the agouti.

It is true that the great continents, such as Asia, Africa, the two Americas, and New Holland, possess large quadrupeds, and, generally speaking, contain species peculiar to each; insomuch that whenever large countries of this description have been discovered, which their situation has kept isolated Iron1 the rest of the world, the class of quadrupeds which they contained has been found entirely different from any that existed elsewhere. Thus, when the Spaniards first penetrated into South America, they did not find a single species of quadruped the same as any of Europe, Asia, or Africa. The puma, the jaguar, the tapir, the cabiai, the lama, the vicuna, the sloths, the armadilloes, the opossums, and the whole tribe of sapajous, were to them entirely new animals, of which they had no idea. Similar circumstances have recurred in our own time, when the coasts of New Holland and the adjacent islands were first explored. The various species of kangaroo, phascolomys, dasyurus, and perameles, the flying phalangers, the ornithorynchi and echidnae, have astonished naturalists by the strangeness of their conformations, which presented proportions contrary to all former rules, and wrere incapable of being arranged under any of the systems then in use.

If there yet remained some great continent to be discovered, we might still hope to become acquainted with new species, among which there might be found some having more or less similarity to those of which we have discovered the remains in the bowels of the earth. But it is sufficient to cast a glance over the map of the world, and see the innumerable directions in which navigators have traversed the ocean, in order to be satisfied that there remains no other large land to be discovered, unless it may be situated towards the South Pole, where the existence of life would necessarily be precluded by the accumulation of ice.

Hence, it is only from the interior of the large divisions of the world, that we can have any hope of still procuring quadrupeds hitherto unknown. But a little reflection will be sufficient to convince us, that our expectations from this source have as little foundation as from that of the islands.

Doubtless, the European traveller cannot easily traverse vast extents of countries, which are either destituteof inhabitants, or are peopled only with ferocious tribes; and this more especially true with regard to Africa. But there is nothing to prevent the animals themselves from roaming over these countries in all directions, and penetrating to the coasts. Even when there may be great chains of mountains between the coasts and the deserts of the interior, they must, always be broken in some places to allow the rivers to pass through; and in these burning deserts, the quadrupeds naturaily follow the banks of rivers. The inhabitants of the coasts also ascend these rivers, and soon become acquainted with all the remarkable species which exist even to their sources, either from personal observation, or by means of intercourse with the inhabitants of the interior. At no period, therefore, could civilized nations have frequented the coast of a large country for any considerable length of time, without gaining some tolerable knowledge of such of the animals which it contained as were remarkable for their size or configuration.

This reasoning is confirmed by well known facts. Although the ancients never passed the mountains of Imaus, or crossed the Ganges, in Asia; and although they never penetrated very far beyond Mount Atlas, a Africa; yet were they, in reality, acquainted with all the large animals of these two divisions of the world; and if they have not distinguished all the species, it was not because they had not seen them, or heard them spoken of by others, but because the mutual resemblances of some of these species caused them to be confounded together. The only important exception which can be opposed to this assertion, presents itself in the Tapir of Malacca, recently sent home from India by two young naturalists, pupils of mine, Messrs Duvaucel and Diard, and which in fact is one of the most interesting discoveries with which Natural History has been enriched m these latter times.

The ancients were perfectly acquainted with the Elephant; and the history of that quadruped is given more accurately by Aristotle than by Buffon. They were not even ignorant of some of the differences which distinguish the elephants of Africa from those of Asia.

They knew the two-horned Rhinoceros, which has never been seen alive in modern Europe. Domitian exhibited it at Rome, and had it stamped on his medals, which have beer very well described by Pausanias.

The one-horned Rhinoceros, distant as was its country, was equally known to them. Pompey shewed one at Rome; and Strabo has accurately described another which he saw at Alexandria. The Rhnoceros of Sumatra described by Mr Bell; and that of Java, discovered and sent home by Messrs Duvaucel and Diard, do not appear to inhabit the continent. Hence, it is not surprising, that the ancients should have been ignorant of them; besides, they probably would not have distinguished them from the others.

The Hippopotamus has not been so well described as the preceding animals; yet very exact representations of it have been left by the Romans in their monuments relative to Egypt, such as the statue of the Nile, the Palestrine pavement, and a great number of medals. In fact, tills animal was repeatedly seen by the Romans; having been exhibited by Scaurus, Augustus, Antoninus, Commodus, Heliogabalus, Philip, and Carinus.

The two species of Camel, the Bactrian and Arabian, are both very well described and characterized by Aristotle.

The Giraffe, or Camelopard (Camel-Leopard), was also well known to the ancients. A live one was shown at Rome, in the circus, during the dictatorship of Julius Caesar, in the year of Rome 708; and ten of them were exhibited together by Gordian III all of which were killed at the secular games of Philip,—a circumstance which may well surprise the moderns, who have only witnessed a single individual, which was sent by the Sultan of Egypt to Laurentius de Medicis, in the fifteenth century, and is painted in the frescoes of Poggio-Cajano.

If we read with attention the descriptions of the Hippopotamus, given by Herodotus aud Aristotle, and which are supposed to have been borrowed from Ilecatams of Miletum, we shall find, that they must have been made up from two different animals, one of which was perhaps the true hippopotamus, and the other was assuredly the Gnou, a quadruped, of which our naturalists begin to take notice only about the end of the eighteenth century. It is the same animal of which fabulous accounts were given by Pliny and Julian, under the name of catoblepas and catablepon.

The Ethiopian Boar of Agatharchides, which is described as having horns, is precisely the Ethiopian Boar of modern times, the enormous tusks of which deserve the name of horns nearly as much as those of the elephant.

The Bubalus and Nagor are described by Pliny; the Gazelle byAelian; the Oryx by Oppian; the Axis, so early as the time of Ctesias; and the Algazel, and Corinne, are accurately figured upon the Egyptian rnonuments.

Julian has well described the Bos grunniens or Yak, under the name of the ox having a tail which serves for a fly-flapper.

The Buffalo was not domesticated by the ancients; but the Indian Ox, of which Julian speaks and which had horns large enough to bold three amphorae, was assuredly that variety of the buffalo which is now called the arnee. And even the wild ox with depressed horns, which is mentioned by Aristotle as inhabiting Arachosia, a province of ancient Persia, could be nothing else than the common buffalo.

The ancients were acquainted with the hornless variety of the ox, and with the African oxen, whose horns, being only attached to the skin, moved with it. They also knew the Indian oxen, which equalled the horse in speed; and those which were so small as not to exceed a he-goat in size. Nor were the broad tailed sheep unknown to them,—nor those of India, which were sa id to be as large as asses.

Although the accounts left us by the ancients, respecting the Aurochs, the rein-deer, and Elk, are all mingled with fable, they are yet sufficient to prove that these animals were in some degree known to them, but that the reports which had reached them, had been communicated by ignorant people, and had not been corrected by a judicious examination. These animals still inhablt the countries which the ancients assigned to themand have only disappearcd in such of them as have been too much cultivated for tlieu habits. The aurochs and elk still exist in the forests of Lithuania, which were formerly continuous with the great Hercynian Forest. The former of these animals still occurs in the northern parts of Greece, as it did in the days of Pausanias. The rein-deer inhibits the snowy regions of the north, where it always had its abode; it changes its colour, not at pleasure, but according to the change of the seasons. It was in consequence of mistakes scarcely excusable, that it was imagined to have occurred in the Pyrenees in the fourteenth century

Even the White Bear had been seen in Egypt white under the Ptolemies .

Lions and Panthers were common at Rome, where they were presented by hundreds in the games of the Circus. Even several Tigers were exhibited there, as well as the Striped Hyena and the Crocodile of the Nile. In the ancient mosaics preserved at Rome, there are excellent representations of the rarest of these animals. Among others, the striped hyena is seen represented with accuracy in a fragment preserved in the Museum of the Vatican; and, while I was at Rome in 1809, a mosaic pavement, composed of natural stones, arranged in the Florentine manner, was discovered in a garden beside the triumphal arch of Galienus, which represented four Bengal tigers executed in a superior manner.

In the Museum of the Vatican, there is deposited the figure of a crocodile in basalt, which is almost a perfect representation of that animal.

It cannot in the least be doubted, that the Hippotigris was the Zebra, which, however, is only found in the southern parts of Africa.

It would be easy to show that almost all the more remarkable species of Apes and Monkeys have been distinctly indicated toy the ancients, under the names of Pitheci, Sphinxes, Satyri, Cebi, Cynocephali, and Cercopitheci.

They even knew, and have described several species of Glires of inconsiderable size, when these animals presented any thing remarkable in their conformation or properties. But the small species are of no importance with reference to the object in view; and, it is sufficient for our purpose to have shewn, that all the large species, which possess any remarkable character, and which we know to inhabit Europe, Asia, and Africa, at the present day, were known to the ancients, whence we may fairly conclude, that their silence in respect to the small quadrupeds, and their neglect in distinguishing the species which very nearly resemble each other, as the various species of antelopes, and of some other genera, were occasioned by want of attention and ignorance of methodical arrangement, rather than by any difficulty proceeding from climate. We may also conclude, with equal certainty, that, as the lapse of eighteen or twenty centuries, together with the advantages of circumnavigating Africa, and of penetrating into India, have added nothing in this department to the information left us by the ancients, there is no probability that succeeding ages will add much to the knowledge of our posterity.

But perhaps some persons may be disposed to employ an opposite train of argument, and to allege that the ancients were not only acquainted with as many large quadupeds as we are, as has already been shown, but that they have described several others which we do not now know— that we act rashly in considering these animals as fabulous,—that we ought to search for them before concluding that we have exhausted the history of the present animal creation,—and in fine, that among those animals which we presume to be fabulous, we may, perhaps, discover, when we become better acquainted with them, the originals of those bones of unknown animals which we discover buried in the earth. Some may even conceive, that those various monsters, which constitute the essential ornaments of the history of the heroic ages of almost all countries, are precisely those very species which it was necessary to destroy, in order to allow the establishment of civilization. Thus the Theseuses and Bellerophons of ancient times had been more fortunate than all the nations of our days, which have only been able to drive back the noxious animals, but have never yet succeeded in exterminating a single species

5.Inquiry respecting the Fabulous Animals of the Ancients.

It is easy to reply to the foregoing objection, by examining the descriptions of these unknown beings, and by inquiring into their origins. The greater number of them have an origin purely mythological, and of this origin their descriptions bear unequivocal marks; for in almost all of them we see merely parts of known animals united by an unbridled imagination, and in contradiction to all the laws of nature.

Those which were invented or arranged by the Greeks, have at least the merit of possessing elegance in then composition Like those arabesques which docorate the remains of some ancient buildings, and which have been multiplied by the fertile pencil of Raphael, the forms which they combine, however repugnant to reason they may be, present agreeable contours. They are the fantastic productions of playful genius; perhaps emblematic representations in the oriental taste, in which were supposed to be concealed under mystical images certain propositions in metaphysics or in morals. We may excuse those who employ their time in attempts to discover the wisdom concealed in the sphinx of Thebes, the pegasus of Thessaly, the minotaur of Crete, or the chimera of Epirus; but it would be absurd to expect seriously to find such productions in nature. As well might we search for the animals described in the Book of Daniel, or for the beast of the Apocalypse.

Neither may we look for the mythological am mals of the Persians, creatures of a still bolder imagination: the martichore, or man-destroyer, bearing a human head on the body of a lion, terminated by the tail of a scorpion; the griffon, guardian of treasures, half eagle, half lion; the cartazonon, or wild ass, armed with a long horn on its forehead .

Ctesias, who has described these as real animals has been looked upon by many authors as an inventor of fables; whereas he has merely attributed an actual existence to emblematical figures These imaginary compositions have been seen m modern times sculptured upon the ruins of Persepolis. What they were intended to signify we shall probably never know; but of this much we are certain, that they do not represent actual beings.

Agatharchidas, another fabricator of animals, drew his information in all probability from a similar source. The ancient Egyptian monuments still furnish us with numerous fantastic representations, in which the parts of different species are combined : gods are often figured with a human body and the head of an animal, and animals are seen with human heads; thus giving rise to the cynocephali, sphinxes, and satyrs of ancient naturalists. The custom of representing in the same painting men of very different sizes, of making the king or the conqueror gigantic, the subjects or the conquered three or four times smaller, must have given rise to the fable of the pigmies. It was in some corner of one of these monuments that Agatharchidas must have seen his carnivorous bull, which with mouth extending from ear to ear devored every other animal. Certainly no naturalist would admit the existence of such an animal; for nature never conbines either cloven hoofs or horns with teeth adapted for devoring animal food.

There may perhaps have been many other figures equally strange, either among such of these monuments as have not been able to resist the ravages of time, or in the temples of Ethiopia and Arabia, which have been destroyed by the religious zeal of the Mahometans and Abyssinians. The monuments of India teem with such figures; but the combinations in these are too extravagant to have deceived any one, Monsters with a hundred arms, and twenty heads all different from one another, are far too absurd to be believed. Nay, the inhabitants of Japan and China also have their imaginary animals, which they represent as real, and which figure even in their religious books. The Mexicans had them. In short, they are the fashion among all nations, whether at the periods when their idolatry has not yet been refined, or when the import of these emblematical combinations has been lost. But who would dare to affirm that he had found those productions of ignorance and superstition in nature? And yet it may have happened that travellers, influenced by a desire of making themselves famous, might pretend that they had seen those strange beings, or that, deceived by a slight resemblance, into which they were too careless to enquire, they may have taken real animals for them. In the eyes of such people, large baboons or monkeys may have appeared true cynocephali, sphinxes, or men with tails. It is thus that St Augustin may have imagined he had seen a satyr.

Some real animals, inaccurately observed and described, may have given rise to monstrous ideas, which, however, have had their foundation in some reality. Thus, we can have no doubt of the existence of the hyena, although that animal has not its neck supported by a single bone and although it does not change its sex every year, as Pliny alleges. Thus, also, the carnivorous bull is perhaps nothing else than a two-horned rhinoceros erroneously described. M. de Weltheim affirms with probability, that the auriferous ants of Herodotus are corsacs.

One of the most famous amongst these fabulous animals of the ancients, is the unicorn. Even to our own time people have obstinately persisted in searching for it, or, at least, in seeking arguments to prove its existence. Three several animals are frequently mentioned by the ancients as having only one horn in the middle of the forehead. The African oryx, having cloven hoofs, the hair placed in the contrary direction to that of other animals, equal in size to the bull or even the rhinoceros and said to resemble deer and goats in form; the Indian ass, having solid hoofs; and the munoceros, properly so called, whose feet are sometimes compared to those of the lion, and sometimes to those of the elephant and which is therefore considered as having divided feet The one-horned horse and one-horned bull are doubtless both to be referred to the Indian ass, for even the latter is described as having solid hoofs. I would ask, If these animals exist as distinct species, should we not at least have their horns in our collections? And what single horns do we possess, excepting those of the rhinoceros and narwal?

How is it possible, after this, to refer to rude figures traced by savages upon rocks? Ignorant of perspective, and wishing to represent a straight horned antelope in profile, they could only give it a single horn, and thus they produced an oryx. The oryxes, too, that are seen on the Egyptian monuments, are probably nothing more than productions of the stiff style, imposed upon the artists of that country by their religion. Many of their profiles of quadrupeds show only one fore and one hind leg; and this being the case, why should they have shown two horns? It may perhaps have chanced that individuals have been taken in the chace, which had accidentally lost one of their horns, as pretty frequently happens to the chamois and saiga: and this would have been sufficient to confirm the error produced by these representations. It is probably in this way that the unicorn has recently been reported to be found in the mountains of Thibet.

All the ancients, however, have not represented the oryx as having only one horn. Oppian expressly gives it several, and Aelian mentions oryxes which had four. Finally, if this animal was ruminant and cloven-hoofed, we know assuredly that its frontal bone must have been longitudinally divided into two, and that it could not, as is very justly remarked by Camper, have had a horn placed upon the suture.

But it may be asked, What two-horned animal could have given the idea of the oryx, and presented the characters which it is described as possessing with regard to its conformation, even independent of the notion of a single liom? To this I reply, with Pallas, that it was the straight horned antelope, the Antilope oryx of Gmelin, improperly named pasan by Buffon. It inhabits the deserts of Africa, and must approach the confines of Egypt. It is this animal which the hieroglyphics appear to represent. Its form is nearly that of the stag; its size equals that of the bull; the hair of its back is directed toward the head; its horns form exceedingly formidable weapons, pointed like javelins, and hard as iron; its hair is whitish, and its face is marked with spots and streaks of black. Such is the description given of it by naturalists; and the fables of the Egyptian priests, which have occasioned the insertion of its figure among their hieroglyphics, do not require to have been founded in nature. Supposing, therefore, that an individual of this species had been seen which had lost one of its horns by some accident, it might have been taken as a representative of the whole race, and erroneously adopted by Aristotle, and copied by his successors. All this is possible, and even natural, and yet proves nothing with regard to the existence of a single-horned species.

In regard to the Indian ass, if we attend to the properties ascribed to its horns as an antidote against poison, we shall see that they are precisely the same as those which the eastern nations attribute at the present day to the horn of the rhinoceros. When this horn was first imported into Greece, the animal to which it belonged might still have been unknown. In fact, Aristotle makes no mention of the rhinoceros, and Agatharchides was the first who described it. In the same manner, ivory was in use among the ancients long before they were acquainted with the elephant. It is even possible that some of their travellers might have given to the rhinoceros the name of Indian ass, with as much propriety as the Romans denominated the elephant the bull of Lucania. Everything, moreover, that is said of the strength, size, and ferocity of this wild ass of theirs, corresponds very well with the rhinoceros. In succeeding times, naturalists, who had now become better acquainted with the rhinoceros, finding tins denomination of Indian ass in the writings of authors who had preceded them, might have taken it, from want of proper examination, for that of a distinct animal; and from the name, they would have concluded the animal should have solid hoofs There is, indeed, a full description of the Indian ass given by Ctcsias, but we have seen above that it had been taken from the bas-reliefs of Persepolis, and must therefore go for nothing in the real history of the animal.

When there afterwards appeared more exact descriptions of an animal having a single horn only, but with several toes, a third species would have been made out, to which they gave the name of monoceros. These double references applied to the same species, are more frequent among ancient naturalists, because most of their works which have come down to us were mere compilations; even because Aristotle himself has frequently mingled facts borrowed from others with those which he had observed himself; and because the habit of critical examination was then as little known among naturalists as among historians.

From all these reasonings and digressions, it may be fairly concluded, that the large animals of the old continent with which we are now acquainted, wcre known to the ancients; and that the animals described by the ancients, and which arc now unknown, were fabulous. It also follows, that the large animals of the three principal parts of the then discovered world could not have been long in being known to the nations which frequented their coasts.

It may also be concluded, that no large species remains to be discovered in America. If there were any, there can be no reason why we should not be acquainted with it; and in fact none has been discovered there during the last hundred and fifty years. The tapir, the jaguar, the puma, the cabiai, the lama, the vicunna, the red wolf, the buffalo or American bison, the ant-eaters, sloths and armadilloes, are as well described by Margrave and Hernandez as by Buffon; it may even be said that they are better, for Buffon has confused the history of the ant-eaters, mistaken the jaguar and red wolf, and confounded the bison of America with the aurochs of Poland. Pennant.

it is true, was the first naturalist who clearly distinguished the small musk ox; but it was long before made mention of by travellers. The cloven-footed horse of Molina, has not been described by the early Spanish travellers; but its existence is more than doubtful, and the authority of Molina is too suspicious to authorise our adopting it. It might be possible to characterise more accurately than has been done the different species of deer belonging to America and India; but the case is with respect to these animals as it was among the ancients with respect to the antelopes; it is the want of a good method for distinguishing them, and not of opportunities of seeing them, that has left them so imperfectly known to us. It may, therefore, be said, that the Mouflon of the Blue Mountains is the only American quadruped of any considerable size of which the discovery is altogether modern; and even it is perhaps only an argali that may have crossed upon the ice from Siberia.

How should it be thought, after this, that the huge mastodons and gigantic megatheria, whose bones have been discovered under ground in North and South America, still exist alive on that continent? How should they have escaped those wandering tribes which continually traverse the country in all directions, and which are themselves aware that these animals no longer exist, since they have invented a fabulous account of their destruction, alleging that they were killed by the Great Spirit, to prevent them from extirpating the human race. But it is evident that this fable has been occasioned by the discovery of the bones, like that of the inhabitants of Siberia with respect to their mammoth, which they pretend to live under ground like the mole, and, like all those of the ancients, about the graves of giants, who were thought by them to have been buried wherever the bones of elephants were discovered.

Thus it may safely be concluded, that if, as we have just said, none of the large species of quadrupeds whose remains are at the present day found in regular mineral strata, bear resemblance to any of the known living species, this is not the effect of mere chance, nor because those species of which we possess nothing but the bones, are still concealed in the deserts, and have hitherto eluded the observation of travellers. On the contrary, this phenomenon must be regarded as resulting from general causes; and its investigation may be considered as affording one of the best means for discovering the nature of these causes.

6.Difficulty of determining the Fossil Bones of Quadrupeds

If this study is more satisfactory in its results than that of other fossil remains of animals, il is also beset with more numerous difficulties. Fossil shells usually present themselves in an entire state, and with all the characters requisite for comparing them with their analogous species, preserved in the collections or figured in the works of naturalists. Even fishes present their skeleton more or less entire; the general form of their body is almost always distinguishable, and most commonly, also, their generic and specific characters, which are drawn from their solid parts. In quadrupeds, on the contrary, even should the skeleton be found entire, it would be difficult to apply to it characters derived; for the most part, from the hair, the colours, and other marks which have disappeared previous to their incrustation. It is even excessively rare to find a fossil skeleton approaching in any considerable degree to a complete state. The strata for the most part, only contain separate bones, scattered confusedly, and almost always broken, and reduced to fragments; and these constitute the only resources of knowledge to the naturalist in this department. It may also be stated, that most observers, deterred by these difficulties, have passed slightly over the fossil bones of quadrupeds; have classed them in a vague manner, according to superficial resemblances, or have not even ventured to assign them a name; so that this part of the history of fossil remains, although the most important and most instructive of all, is, at the same time, that which has been the least cultivated.

Principle by which this determination is effected

Fortunately, comparative anatomy possesses a principle, which, when properly developed, enables us to surmount all the obstacles. This principle consists in the mutual relation of forms in organised beings, by means of which, each species may be determined, with perfect certainty, by any fragment of any of its parts.

Every organised being forms a whole,—a peculiar system of its own, the parts of which mutually correspond, and concur in producing the same definitive action, by a reciprocal reaction. None of these parts can change in form, without the others also changing; and consequently, each of them, taken separately, indicates and ascertains all the others.

Thus, if the intestines of an animal are so organised as to be fitted for the digestion of flesh only, and that flesh recent, it is necessary that its jaws be so constructed as to fit them for devouring live prey; its claws for seeing and tearing it; its teeth for cutting and dividing it; the whole system of its organs of motion, for pursuing and overtaking it; and its organs of sense for discovering it at a distance. It is even requisite that nature have placed in its brain the instinct necessary for teaching it to conceal itself, and to lay snares for its victims.

Such are the general conditions which nature imposes upon the structure of carnivorous animals: and which every animal of this description must indispensably combine in its constitution, for vithout them its race could not subsist. But subordinate to these general conditions, there exist others, having relation to the size, the species, and the haunts of the prey for which the animal is adapted; and from each of these particular conditions, there result modifications of detail in the forms which arise from the general conditions.

Thus not only the class, but the order, the genus, and even the species, are found expressed in the form of each part.

In fact, in order that the jaw may be able to seize, it must have a certain form of condyle; that the resistance, the moving power, and the fulcrum, should have a certain relative position in regard to each other; and that the temporal muscles should be of a certain size; the hollow or depression, too, in which these muscles are lodged, must have a certain depth; and the zygomatic arch, under which they pass, must not only have a certain degree of convexity, but it must be sufficiently strong to support the action of the masseter.

In order that the animal may be able to carry off its prey, it must have a certain degree of vigour in the muscles which elevate the head; whence there results a determinate form in the vertebrae from which these muscles take their rise, and in the occiput into which they are inserted.

In order that the teeth may be able to cut flesh, they must he sharp-edged, and must bc so in a greater or less degree, according as they have flesh more or less exclusively to cut. Their base will be solid, according to the quantity and size of the bones which they have to break. The whole of these circumstances must necessarily influence the development and form of all the parts which contribute to move the jaws.

In order that the paws may be able to seize the prey, there must be a certain degree of mobility in the toes, and a certain degree of strength in the claws, from which there will result determinate forms in all the phalanges, and a corresponding distribution of muscles and tendons. The fore-arm, or cubitus, must possess a certain facility of turning, from which there will also result determinate forms in the bones of which it is composed But the bones of the cubitus being articulated to the humerus, a change in the proportions of the former, will necessarily induce a corresponding change in the latter. The shoulder-bones must have a certain degree of firmness in such animals as make use of their forelegs for seizing, and from this there must also result a certain peculiarity in their form. The play of all these parts will require certain proportions in all their muscles, and the impressions made by these muscles so proportioned, will determine still more particularly the forms of the bones.

It is easy to see that similar conclusions may be drawn with regard to the posterior extremities which contribute to the rapidity of the general motions; with regard to the composition of the trunk, and the forms of the vertebra, which exert an influence upon the facility and flexibility of these motions; and, lastly, with regard to the forms of the bones of the nose, of the orbit, and of the ear, the connection of which with the perfection of the senses of smell, sight, and hearing, is evident. In a word, the form of the tooth regulates the forms of the condyle, of the scapula, and of the claws, in the same manner as the equation of a curve regulates all its properties; and as, by taking each property separately for the base of a particular equation, we find both the ordinary equation, and all the other properties whatever; so, the claw, the scapula, the condyle, the femur, and all the other bones taken separately, give the tooth, or are reciprocally given by it; and thus, by commencing with any one of these bones, a person who possesses an accurate knowledge of the laws of organic economy, may reconstruct the whole animal.

This principle seems sufficiently evident, in the general acceptation in which it is here taken, and does not require any fuller demonstration, but when it comes to be applied, there will be found many cases where our theoretical knowledge of the relations of forms will not be sufficient, unless it be supported by observation and experience.

For example, we are well aware, that hoofed animals must all be herbivorous, since they have no means of seizing prey. It is also evident, that, having no other use to make of then forelegs than to support their body, they do not require a shoulder so vigorously organised as that of carnivorous animals; they have, therefore, no acromion or clavicle, and their shoulder-blades are narrow. Having also no occasion to turn their fore-arms, their radius is united to the ulna by ossification, or at least articulated by a ginglimus or hinge-joint, and not by arthrodia or ball and socket, to the humerus. Their food being herbaceous, will require teeth furnished with flat surfaces, for bruising seeds and plants. The crown of the teeth must also be unequal, and, for this purpose, must be composed of parts alternately consisting of bone and of enamel. Teeth of this structure necessarily require horizontal motions to enable them to triturate the food; and hence the condyle of the jaw cannot be so strictly confined within its articulating cavity as in the carnivorous animals, but must be flattened, and thus correspond with a more or less flattened surface of the temporal bones. Further, the temporal fossa, which will only have a small muscle to contain, will be narrower, and not so shallow, as that of carnivorous animals. All these circumstances are deductible from each other, according to their greater or less generality, and in such a manner, that some of them are essential and exclusively peculiar to hoofed animals, while others, although equally necessary in these animals, are not entirely peculiar to them, but may occur in other animals also, where the rest of the conditions will permit their existence.

If we proceed to consider the orders or subdivisions of the class of hoofed animals, and examine what modifications the general conditions undergo, or rather what particular conditions are conjoined with them, according to the respective characters of these orders, the reasons of these subordinate conditions begin to appear less obvious. We can still easily conceive, in general, the necessity of a more complicated system of digestive organs in those species which have a more imperfect masticatory system; and hence we may presume, that these latter must be rather ruminating animals, in which there is wanting such or such an order of teeth; and may also deduce from the same consideration, the necessity of a certain form of the oesophagus, and of corresponding forms in the vertebra; of the neck, &c. But I doubt whether it would have been discovered, independently of actual observation, that the ruminating animals should all have cloven hoofs, and that they should be the only animals having them; that there should be horns on the forehead in this class alone or that such of them as have sharp canine teeth, should, in general, have no horns.

However, since these relations are constant, we may be assured that they have a sufficient cause; but as we are not acquainted with that cause, we must supply the defect of theory by means of observation, and in this way establish empirical laws which become nearly as certain as those deduced from rational principles, when founded upon observations, the authenticity of which is proved by frequent repetition. Hence, at the present day, any one who observes only the print of a cloven foot, may conclude that the animal which left this impression ruminates; and this conclusion is quite as certain as any other in physics, or in moral philosophy. This simple footmark, therefore, indicates at once to the observer the forms of the teeth, of the laws, of the vertebrae, of all the bones of the legs, thighs, shoulders, and pelvis of the animal which had passed. It is a surer mark than all those of Zadig. That there are secret reasons, however, for all these relations, is what observation alone is sufficient to show, ndependently of any general principles of philosophy.

In fact, when we construct a table of these relations, we remark not only a specific constancy. If the expression may be allowed, between a particular form of a particular organ, and some other form of a different organ; but we also perceive a classic constancy of conformation, and a corresponding gradation, in the development of these two organs, which demonstrate their mutual influence, almost as well as the most perfect deduction of reason.

For example, the dentary system of the hoofed animals, which are not ruminant, is in general more perfect than that of the cloven-footed or ruminating animals, because the former have either incisors, or canine teeth, and almost always both in each jaw; and the structure of their foot is in general more complicated, because they have more toes or claws, or their phalanges less enveloped in the hoof,—or a greater number of distinct bones in the metacarpus and metatarsus—or more numerous tarsal bones—or a fibula more distinct from the tibia—or, lastly, that all these circumstances are often united in the same species of animals.

It is impossible to assign reasons for these rela tions; but we are certain that they are not the effects of chance, because, whenever a cloven-footed animal manifests, in the arrangement of its teeth some tendency to approach the animals we now speak of, it also manifests a similar tendency in the arrangement of its feet. Thus the camels, which have canine teeth, and even two or four incisors in the upper jaw, have an additional bone in the tarsus, because their scaphoid bone is not united to the cuboid, and they have very small hoofs, with corresponding phalanges. The musk animals, whose canine teeth are much developed, have a distinct fibula along the whole length of their tibia; while the other cloven-footed animals have only, in place of a fibula, a small bone articulated at the lower end of the tibia : here is, therefore, a constant harmony between two organs apparently having no connection; and the gradations of their forms preserve an uninterrupted correspondence, even in those cases in which we cannot account for their relations.

Now, by thus adopting the method of observation as a supplementary means, when theory is no longer able to direct our views, we arrive at astonishing results. The smallest articulating surface of bone, or the smallest apophysis, has a determinate character, relative to the class, the order, the genus, and the species to which it belonged; insomuch, that when one possesses merely a well preserved extremity of a bone, he can, by careful examination, and the aid of a tolerable analogical knowledge, and of accurate comparison, determine all these things with as much certainty as if he had the entire animal before him. I have often made trial of this method upon portions of known animals, before reposing full confidence upon it, in regard to fossil remains; and it has always proved so completely satisfactory, that I have no longer any doubts regarding the certainty of the results which it has afforded me.

It is true, that I have enjoyed all the advantages which were necessary for the undertakirg; and that my favourable situation, in the Museum of Natural History at Paris, and assiduous research for nearly thirty years, have procured me skeletons of all the genera and sub-genera of quadrupeds, and even of many species in some genera, and of several varieties of some species. With such means, it was easy for me to multiply my comparisons, and to verify in all their details the applications which I have made of the various laws deducible from such circumstances as have been stated.

We cannot here enter into a more lengthened detail of this method, and must refer to the large work on Comparative Anatomy, in which all its rules will be found. In the mean time, an intelligent reader may gather a great number of these from the work upon Fossil Bones, if he take the trouble of attending to all the applications of them which we have there made. He will see, that it is by this method alone that we are guided, and that it has almost always sufficed for referring each bone to its species, when it was a living species—to its genus, when it was an unknown species—to its order, when it was a new genus— and to its class, when it belonged to an order not hitherto established—and to assign it, in the three last cases, the proper characters for distinguishing it from the nearest resembling orders, genera, and species. Before the commencement of our researches, naturalists had done no more than this with regard to animals, which they had the opportunity of examining in their entire state. Yet, in this manner, we have determined and classed the remains of more than a hundred and fifty mammiferous and oviparous quadrupeds

View of the General Results of these Researches.

Considered with regard to species, upwards of ninety of these animals are most assuredly hitherto unknown to naturalists; eleven or twelve have so perfect a resemblance to species already known, that the slightest doubt cannot be entertained of their identity; the others exhibit many traits of resemblance to known species, but their comparison has not yet been made with sufficient precision to remove all doubts.

Considered with regard to genera, of the ninety hitherto unknown species, there are nearly sixty that belong to new genera. The other specics rank under genera or subgenera already known.

It may not be without use, also, to consider these animals with regard to the classes and orders to which they belong. Of the hundred and fifty species, about a fourth part are oviparous quadrupeds, and all the rest mammifera. Of these last, more than the half belong to non-ruminant hoofed animals.

Notwithstanding what has been done, it would still be premature to establish upon these numbers any conclusion relative to the theory of the earth, because they are not in sufficient proportion to the numbers of genera and species which may be buried in the strata of the earth. Hitherto the bones of the larger species have been chiefly collected, those being more obvious to agricultural labourers; while the bones of the smaller species are usually neglected, unless when they chance to fall into the hands of a naturalist, or when some particular circumstance, such as then excessive abundance in certain places, attracts the attention even of the common people.

7.Relations of the Species of Fossil Animals with the Strata in which they are found.

The most important consideration, that which, in fact, is the chief object of all my researches, and which establishes their legitimate connection with the Theory of the Earth, is to ascertain in what strata each species is found, and whether there may be some general laws, relative either to the zoological subdivisions, or to the greater or less resemblance of the species to those of the present day.

The laws which have been recognised with respect to these relations are very distinct and satisfactory

In the first place, it is clearly ascertained that the oviparous quadrupeds appear much more early than the viviparous; that they are even more abundant, larger, and more varied, in the ancient strata than at the surface of the globe, as it exists at present

The Ichthyosauri, the Plesicsauri, several species of Tortoise, and several species of Crocodile, are found beneath the chalk, in the deposits commonly called Jura formations. The Monitors of Thuringia would be still older, if, according to the Wernerian School, the copper-slate in which they are contained, along with a great variety of fishes supposed to have belonged to fresh-water, is to be placed among the oldest beds of the se-ondary formations. The enormous crocodiles and the great tortoises of Maastricht, are found in the chalk formation itself; but these are marine animals.

This earliest appearance of fossil bones seems, therefore, already to indicate, that dry lands and fresh waters had existed before the formation of the chalk deposits. But neither at this period, nor while the chalk was forming, nor even long after, have any bones of land-mammilera been encrusted : or, at least, the small number of these, which are alleged to have been found in strata of these dates, forms but a trifling exception.

We begin to find bones of marine mammiifera, namely, of lamantins and seals, in the coarse shelly limestone which covers the chalk in the neighbourhood of Paris; but there are still no bones of terrestrial mammifera.

Notwithstanding the most assiduous investigation, I have not been able to discover any distinct trace of this class in any of the deposits preceding those which rest upon the coarse limestone. Certain lignites and molasses do in fact contain them; but I am very doubtful whether these deposits are all, as is commonly supposed, anterior to that limestone. The places where these bones have been found are so limited, both in extent and in number, as to induce us to suppose some irregularity, or some repetition of the formation containing them. On the contrary, the moment we arrive at the deposits which rest upon the coarse limestone, the bones of land-animals present themselves in great abundance.

As it is reasonable to believe that shells and fishes did not exist at the period of the formation of primitive rocks, we are also led to conclude that the oviparous quadrupeds began to exist along with the fishes, and at the commencement of the period during which the secondary rocks were formed: but that the land-quadrupeds did not appear upon the earth, at least in any considerable number, till long after, and until the coarse limestone strata, which contain the greater number of our genera of shells, although of species different from ours, had been deposited.

It is remarkable that those coarse limestone strata which are used at Paris for building, are the last formed strata which indicate a long and quiet continuance of the sea upon our continents. Above them, indeed, there are found formations containing shells and other marine productions; but these consist of collections of transported matters, sand, marls, sandstones, and clays, which rather indicate transportations that have taken place with more or less violence, than strata formed by tranquil deposition; and, if there be some rocky and regular strata of pretty considerable magnitude, beneath or above these transported matters, they generally exhibit indications of having been deposited from fresh water.

Almost all the known bones of viviparous quadrupeds, therefore, have been found either in those fresh-water formations, or in the allmial formations; and consequently there is every reason to conclude that these quadrupeds have only begun to exist, or, at least, to leave their remains in the strata of our earth, after the last retreat of the sea but one, and during the state of things that preceded its last irruption

But there is also an order in the disposition of these bones with regard to each other; and this order further announces a very remarkable succession in the appearanee of the different species. All the genera which are now unknown, the Palaeotheria, Anaplotheria, &c.. with the position of which we are thoroughly acquainted, belong to the oldest of the formations of which we are now speaking, those which rest immediately upon the coarse limestone. It is chiefly these genera which occupy the regular beds that have been deposited from fresh-water, or certain alluvial beds of very ancient formation, generally composed of sand and rolled pebbles, and which were perhaps the earliest alluvium of that ancient world. Along with these there are also found some lost species of known genera, but in small numbers, and some oviparous quadrupeds and fishes, which appear to have been all inhabitants of fresh-water. The beds which contain them are always more or less covered by alluvial beds, containing shells, and other marine productions. The most celebrated of the unknown species, which belong to known genera, or to genera closely allied to those which are known, such as the fossil elephants, rhinoceroses, hippopotami, and mastodons, do not occur along with those more ancient genera. It is in the alluvial formations alone that they are discovered, sometimes accompanied with marine shells, and sometimes with fresh-water shells, but never in regular stony beds. Every thing that is found along with these species is either unknown like themselves, or at least doubtful.

Lastly, the bones of species which are apparently the same as those that are still found alive, are never discovered, except in the last alluvial deposits formed on the sides of rivers, or on the bottoms of ancient pools or marshes now dried up, or in the substance of beds of peat, or in the fissures and caverns of some rocks; or, lastly, at small depths below the surface, in places where they may have been buried by the falling down of debris, or even by the hand of man: and their superficial position renders these bones, although the most recent of all, almost always the worst preserved.

It must not, however, be thought that this classification of the various geological positions of fossil remains, is as certain as that of the species, or that it is equally capable of demonstration. There are numerous reasons which prevent this from being the case.

In the first place, all my determinations of species have been made upon the bones themselves, or by means of good figures; whereas it has been impossible for me personally to examine all the places in which these bones have been discovered. I have very frequently been obliged to content myself with vague and ambiguous accounts, given by people who were not themselves well aware of what it was necessary to observe; and, more fre-quently still, I have been unable to procure any information whatever on the subject.

Secondly, these repositories ot organic remains are subject to infinitely greater doubts, than the bones themselves. The same formation may appear recent in places where it shews itself at the surface, and ancient in those where it is covered by the beds which have succeeded it. Ancient formations may have been transported by partial inundations, and thus have covered recent bones; they may have fallen upon them by crumbling, and thus have enveloped and mingled them with the productions of the ancient sea, which they previously contained. Bones of ancient periods may have been washed out by the waters, and afterwards enveloped in recent alluvial formations. Lastly, recent bones may have fallen onto the fissures of caverns of ancient rocks, and been enveloped by stalactites or other incrustations. In every individual instance, therefore, it becomes necessary to analyze and appreciate all those circumstances which might disguise the real origin of fossil remains; and it rarely happens that people who have collected bones have been themselves aware of this necessity, the consequence of which has been, that the true characters of their geological position have been almost always neglected or misunderstood. Thirdly, there are some doubtful species, which must occasion more or less uncertainty in the results of our researches, until they have been clearly ascertained. Thus the horses and buffaloes that occur along with the elephants, have not yet received appropiate specific characters; and such geologists as are disinclined to adopt the different, epochs which I have endeavoured to establish with regard to fossil bones, may, for many years to come, draw from thence an argument against my system, so much the more convenient as it is contained in my own work. But allowing that these epochs are liable to some objections, from such as may only consider some particular case, I am not the less satisfied, that those who shall take a comprehensive view of the phenomena, will not be cheeked by such inconsiderable and partial difficulties, and will be led to conclude, as I have done, that there has been at least one, and very probably two, successions in the class of quadrupeds, previous to that which at the present day peoples the surface of the earth.

Proofs that the Extinct Species of Quadrupeds are not varieties of the presently existing Species. I now proceed to the consideration of another objection, one, in fact, which has already been urged against me.

Why may not the presently existing races of land quadrupeds, it has been asked, be modifications of those ancient races which we find in a fossil state; which modifications may have been produced by local circumstances and change of climate, and carried to the extreme difference which they now present, during a long succession of ages?

This objection must appear strong to those especially who believe in the possibility of indefinite alteration of forms in organised bodies; and who think that, during a succession of ages, and by repeated changes of habitudes, all the species might be changed into one another, or might result from a single species.

Yet to these persons an answer may be given from their own system. If the species have changed by degrees, we ought to find traces of these gradual modifications. Thus, between the paleotheria and our present species, we should be able to discover some intermediate forms; and yet no such discovery has ever been made.

Why have not the bowels of the earth preserved the monuments of so strange a genealogy, if t be not because the species of former times were as constant as ours or, at least, because the catastrophe which destroyed them, had not left them sufficient time for undergoing the variation alleged?

In order to reply to those naturalists who acknowledge that the varieties of animals are restrained within certain limits fixed by nature, it would be necessary to examine how far these limits extend This is a very curious inquiry,—highly interesting in itself, under a variety of relations, and yet one that has been hitherto very little attended to.

Before entering upon this inquiry, it is proper to define what is meant by a species, so that the definition may serve to regulate the employment of the term. A species, therefore, may be defined, as comprehending the individuals which descend from each other, or from common parents, and those which resemble them as much as they resemble each other. Thus, we consider as varieties of a species, only the races more or less different which may have sprung from it by generation. Our observations, therefore, regarding the differences between the ancestors and descendants, afford us the only certain rule by which we can judge on this subject; all other considerations leading to hypothetical conclusions destitute of proof. Now, considering the varieties in this view, we observe that the differences which constitute it, depend upon determinate circumstances, and that their extent increases in proportion to the intensity of these circumstances.

Thus, the most superficial characters are the most variable: the colour depends much upon the light; the thickness of the fur upon the heat; the size, upon the abundance of food. But in a wild animal, even these varieties are greatly limited by the natural habits of the animal itself, which does not voluntarily remove far from the places where it finds, in the necessary degree, all that is requisite, for the support of its species, and does not even extend its haunts to any great distance, unless it also finds all these circumstances conjoined. Thus, although the wolf and the fox inhabit all the climates from the torrid to the frigid zone, we hardly find any other difference among them, in the whole of that vast space, than a little more or a little less beauty in their fur. I have compared skulls of foxes from the northern countries and from Egypt, with those of the foxes of France, and have found no difference but such as might be expected in different individuals. Such of the wild animals as are confined within narrower limits, vary still less, especially those which arc carnivorous. The only difference between the hyena of Persia and that of Morocco, consists in a thicker or a thinner mane.

The wild herbivorous animals feel the influence of climate somewhat more extensively, because there is added to it in their case, the influence of the food, which may happen to differ both as to quantity and quality. Thus, the elephants of one forest are often larger than those of another; and their tusks are somewhat longer in places where their food may happen to be more favourable for the production of the matter of ivory. The same may take place with regard to the horns of rein-deer and stags. But let us compare two elephants the most dissimilar, and we shall not discover the slightest difference in the number and articulations of the bones, the structure of the teeth, &c.

Besides, the herbivorous species, in the wild state, seem more restrained from dispersing than the carnivorous animals, because the sort of food which they require, combines with the temperature to prevent them.

Nature also takes care to guard against the alteration of the species, which might result from their mixture, by the mutual aversion with which it has inspired them. It requires all the ingenuity and all the power of man to accomplish these unions, even between species that have the nearest resemblances. And, when the individuals produced by these forced conjunctions are fruitful, which s very seldom the case, their fecundity does not continue beyond a few generations; and would not probably proceed so far, without a continuance of the same cares which excited it at first. Thus, we never see in our woods individuals intermediate between the hare and the rabbit; between the stag and the doe; or between the martin and the pole-cat.

But the power of man changes this order; it discloses all those variations, of which the type of each species is susceptible; and from them derives productions which the species, if left to themselves, would never have yielded.

Here the degree of the variations is still proportional to the intensity of their cause, which is slavery. It is not very high in the semi-domesticated species, such as the cat. A softer fur; more brilliant colours; greater or less size; these form the whole extent of the variations in this species; for the skeleton of an Angora cat differs in no regular and constant circumstance from that of a wild cat.

In the domesticated herbivorous animals, which we transport into all kinds of climates, and subject to all kinds of management, both with regard to labour and nourishment, we certainly obtain greater variations; but still they are all merely superficial. Greater or less size; longer or shorter horns, or even the want of these entirely; a hump of fat, larger or smaller, on the shoulder; these form the differences between the various races of the common ox or bull: and these differences continue long, even in such breeds as have been transported from the countries in which they were produced, when proper care is taken to prevent crossing.

Of this nature are also the innumerable varieties of the common sheep, which consist chiefly in differences of their fleeces, as the wool which they produce is an important object of attention. These varieties, although not quite so perceptible, are yet sufficiently marked among horses. In general, the forms of the bones vary little; their connections and articulations, and the forms of the large-grinding teeth, never vary at all.

The small size of the tusks in the domestic hog, compared with the wild boar's, and the junction of its cloven hoofs into one in some races, form the extreme point of the differences which we have produced in the domesticated herbivorous quadrupeds.

The most remarkable effects of the influence of man are manifested in the animal which he has reduced most completely under subjection the dog,—that species so entirely devoted to ours, that even the individuals of it seem to have sacrificed to us their will, their interest, and inclination. Transported by man into every part of the world, subjected to all the causes capable of influencing their development, regulated in their sexual intercourse by the pleasure of their masters, dogs vary in colour; in the quantity of their hair, which they sometimes even lose altogether, and in its nature; in size, which varies as one to five in the linear dimesions, amounting to more than a hundred fold in bulk; in the form of the ears, nose, aud tail; in the proportional length of the legs ; in the progressive development of the brain in the domestic varieties, whence results the form of their head, which is sometimes slender, with a lengthened muzzle and flat forehead, and sometimes having a short muzzle and a protuberant forehead : insomuch that the apparent differences between a masfiff and a water-spaniel, and between a greyhound and a pug, are more striking than those that exist between any two species of the same natural genus in a wild state. Finally, and this may be considered as the maximum of variation hitherto known in the animal kingdom, there are races of dogs which have an additional toe on the hind foot, with corresponding tarsal bone; as there are, in the human species, some families that have six fingers on each hand. Yet, in all these varieties, the relations of the bones remain the same, nor does the form of the teeth even change in any perceptible degree . the only variationi n respect to these latter being, that, in some individuals, one additional false grinder appears, some hues on the one side, and sometimes on the other.

Animals, therefore, have natural characters, which resist every kind of influence, whether natural or produced by human interference, and nothing indicates that, with regard to them, time has more effect than climate and domestication.

I am aware that some naturalists lay great stress upon the thousands of ages which they call into action by a dash of the pen; but, in such matters, we can only judge of what a long period of time might produce, by multiplying in idea what a less time produces. With this view, have endeavoured to collect the most ancient documents relating to the forms of animals; and there are none which equal, either in antiquity or abundance, those that Egypt furnishes. It affords us, not only representations of animals, but even their identical bodies embalmed in its catacombs.

I have examined with the greatest attention the figures of quadrupeds and birds sculptured upon the numerous obelisks brought from Egypt to ancient Rome. All these figures possess, in their general character, which alone could be the object of attention to an artist, a perfect resemblance to the species represented, such as we see them at the present day.

On examining the copies made by Kirker and Zoega, we find that, without preserving every trait of the originals in its perfect purity, they have given figures which are easily recognised. We readily distinguish the ibis, the vulture, the owl, the falcon, the Egyptian goose, the lapwing, the landrail, the aspic, the cerastes, the Egyptian hare with its long ears, and even the hippopotamus; and, among the numerous monuments engraved in the great work on Egypt, we sometimes observe the rarest animals, the algazel, for example, which was not seen in Europe until within these few years.

My learned colleague, M. Geoffrey Saint-Hilaire, strongly convinced of the importance of this research, carefully collected in the tombs and temples of Upper and Lower Egypt as many mummies of animals as he could. He has brought home cats, ibises, birds of prey, dogs, monkeys, crocodiles, and the head of an ox, in this state; and there is certainly no more difference to be perceived between these mummies and the species of the same kind now alive, than between the human mummies and the skeletons of men of the present day. A difference may, indeed, be found between the mummies of the ibis and the bird which naturalists have hitherto described under that name; but I have cleared up all doubts on this matter, in a Memoir upon the Ibis, which will be found at the end of this Essay, and in which I have shown that it is still at the present day the same as it was in the time of the Pharaohs. I am aware that, in these, I only cite the monuments of two or three thousand years; but this is the most remote antiquity to which we can resort in such a case.

There is nothing, therefore, to be derived from all the facts hitherto known, that could, in the lightest degree, give support to the opinion that the new genera which I have discovered or established among the fossil remains of animals, anymore than those which have in like manner been discovered or established by other naturalists, the palaeotheria, anoplotheria, megalonyces, mastodonta, pterodactyli, ichthyosauri, &c. might have been the sources of the present race of animals, which have only differed from them through the influence of time or climate. Even if it should prove true, which I am far from believing to be the case, that the fosil elephants, rhinoceroses, elks, and bears, differ no more from those at present existing, than the present races of dogs differ from one another, this would not furnish a sufficient reason for inferring the general dentity of the species, because the races of dogs have been subjected to the influence of domestication, which these other animals neither did nor could experience.

Farther, when I maintain that the rocky beds contain the bones of several genera, and the alluvial strata those of several species which no longer exist, I do not assert that a new creation was required for producing the species existing at the present day. I only say chat they did not originally inhabit the places where we find them at present, and that they must have come from some other part of the globe.

Let us suppose, for instance, that a great eruption of the sea were now to cover the continent of New Holland with a coat of sand or other debris; it would bury the carcases of animals belonging to the genera Kangurus, Phascolomys, Dasyurus, Perameles, flying phalanger, echidna, and ornithorynchus, and it would entirely destroy the species of all these genera, since none of them exist now in any other country.

Were the same revolution to lay dry the numerous narrow straits which separate New Holland from the continent of Asia, it would open a road to the elephants, rhinoceroses, buffaloes, horses, camels, and tigers, and to all the other Asiatic quadrupeds, which would come to people a land where they had been previously unknown.

Were some future naturalist, after having made himself well aquainted with this new race of animals, to search below the surface on which they live, he would find remains of quite a different nature.

What New Holland would be, under the circumstances which we have supposed, Europe, Siberia, and a large portion of America, now actually are. And, perhaps, when other countries shall have been examined, and New Holland among the rest, it will one day be found that they have experienced similar revolutions, I might almost say, mutual changes, of productions. For, if we push the supposition farther, and, after the supply of Asiatic animals to New Holland, admit a second revolution, which destroyed Asia, their original country, those naturalists who might observe them in New Holland, their second country, would be equally at a loss to know whence they had come, as we now are to find out the origin of the races of animals that nhabit our own countries.

I now proceed to apply this manner of reasoning to the human species

Proofs that there are no Fossil Human Bones

It is certain that no human bones have yet been found among fossil remains; and this furnishes an additional proof that the fossil races were not mere varieties of known species, since they could not have been subjected to human influence.

When I assert that human bones have never been found among fossil organic remains, (I must be understood to speak of fossils or petrifactions, properly so called), or, in other words, in the regular strata of the surface of the globe; for in peat-bogs (tourbières), and alluvial deposits, as in burying-grounds, human bones might as well be found as bones of horses, or other common species They might equally be found in fissures of rocks, and in caverns, where they may have been covered over by stalactite; but in the beds which contain the ancient races, among the palaeotheria, and even among the elephants and rhinoceroses, the smallest portion of a human bone has never been discovered. Many of the labourers in the gypsum quarries about Paris, believe that the bones which occur so abundantly in them, are in a great part human; but I have seen several thousands of these bones, and I may safely affirm that not one of them has ever belonged to our species. I have examined at Pavia the groups of bones brought by Spallanzani from the Island of Cerigo; and, notwithstanding the assertion of that celebrated observer, I equally affirm, that there is not one among them that could be shown to be human The homo diluvii testis of Scheuchzer has been restored, in my first edition, to its true genus, which is that of the salamanders; and, in a more recent examination of it at Haarlem, allowed me by the politeness of Mr Van Marum, who permitted me to uncover the parts enveloped in the stone, I obtained complete proof of what I had before announced. Among the bones found at Canstadt, the fragment of a jaw, and some articles of human manufacture, were found; but it is known that the ground was dug up without any precaution, and that no notes were taken of the different depths at which each article was discovered. Every where else, the fragments of bone alleged to be human, are found, on examination, to belong to some animal, whether these fragments have been examined themselves, or merely through the medium of figures. Very recently, some were pretended to have been discovered at Marseilles, in a quarry that had been long neglected; but they have turned out to be impressions of tuyaux marines. Such real human bones as have been exhibited as fossil, belonged to bodies that had fallen into fissures, or had been left in the old galleries of mines, or that had been incrusted; and I extend this assertion even to the human skeletons discovered at Guadaloupe, in a rock formed of fragments of madrepore, thrown up by the sea, and united by water impregnated with calcareous matter. The human hones found near Koestriz, and pointed out by M. de Schlothfim, had been announced as taken out of very old beds; but this estimable naturalist is anxious to make known how much this assertion is still subject to doubt. The same has been the case with the articles of human fabrication. The pieces of iron found at Montmartre are fragments of the tools which the workmen use for putting in blasts of gunpowder, and which sometimes break in the stone.

Yet human bones preserve equally well with those of animals, when placed in the same circumstances. In Egypt, no difference is remarked between the mummies of men and those of quadrupeds. I picked up, from the excavations made some years ago in the ancient church of St Genevieve, human bones that had been interred below the first race, which may even have belonged to some princes of the family of Clovis, and which still retained their forms very perfectly. We do not find in ancient fields of battle that the skeletons of men are more wasted than those of horses, except in so far as they may have been influenced by size; and we find among fossil remains the bones of animals as small as rats, still perfectly preserved.

Every circumstance, therefore, leads to the conclusion, that the human species did not exist in the countries in which the fossil bones have been discovered, at the epoch of the revolutions by which these bones were covered up; for there cannot be a single reason assigned,why men should have entirely escaped from such general catastrophes, or why their remains should not be now found like those of other animals. I do not presume, however, to conclude that man did not exist at all before this epoch. He might then have inhabited some narrow regions, whence he might have repeopled the earth after those terrible events. Perhaps also, the places which he inhabited may have been entirely swallowed up in the abyss, and his bones buried at the bottom of the present seas, with the exception of a small number of individuals, which have continued the species.

However this may be, the establishment of man in those countries in which we have said that the fossil remains of land animals are found, that is to say, in the greatest part of Europe, Asia, and America, has necessarily been posterior, not only to the revolutions which have covered up these bones, but also to those which have laid bare the strata containing them, and which are the last that the globe has undergone. Hence it clearly appears, that no argument in favour of the antiquity of the human species in these different countries can be derived either from those bones themselves, or from the more or less considerable masses of rocks or of earthy materials by which they are covered.

8.Physical Proofs of the Newness of the Present Continents.

On the contrary, by a careful examination of what has taken place on the surface of the globe, since it has been laid dry for the last time, and its continents have assumed their present form, at least in the parts that are somewhat elevated, it may be clearly seen that this last revolution, and consequently the establishment of our existing societies, could not have been very ancient. This result is one of the best established, and, at the same time, one of the least attended to in rational geology; and it is so much the more valuable, that it connects natural and civil history in one uninterrupted series.

When we measure the effects produced in a given time by causes still acting, and compare them with those which the same causes have produced since they have begun to act, we are enabled to determine nearly the instant at which their action commenced; which is necessarily the same as that in which our continents assumed their present form, or that of the last sudden retreat of the waters.

It must, in fact, have been since this last retreat of the waters, that our present steep declivities have begun to disintegrate, and to form heaps of debris at their bases; that our present rivers have begun to flow, and to deposit their alluvial matters; that our present vegetation has begun to extend itself, and to produce soil; that our present cliffs have begun to be corroded by the sea that our present downs have begun to be thrown up by the wind: just as it must have been since this same epoch, that colonies of men have begun, for the first or second time, to spread themselves, and to form establishments in places fitted by nature for their reception. I do not here take the action of volcanoes into account, not only because of the irregularity of their eruptions, but because we have no proofs of their not having been able to exist under the sea; and because, on that account, they cannot serve us as a measure of the time which has elapsed since its last retreat.

Additions of Land by the Action of Rivers.

MM. Deluc and Dolomieu have most carefully examined the progress of the formation of new ground by means of matters washed down by rivers; and although exceedingly opposed to each other on many points of the Theory of the Earth, they agree in this. These formations augment very rapidly; they must have increased still more rapidly at first, when the mountains furnished more materials to the rivers, and yet their extent is still inconsiderable.

Dolomieu's Memoir respecting Egypt tends to prove, that the tongue of land on which Alexander caused his city to be built, did not exist in the days of Homer; that they were then able to navigate directly from the island of Pharos into the gulf afterwards called Lake Mareotis; and that this gulf was then, as indicated by Mene- laus, from fifteen to twenty leagues in length. It had, therefore, only required the nine hundred years that elapsed between the time of Homer and that of Strabo, to bring things to the state in which this latter author describes them, and to reduce the gulf in question to the form of a lake, of six leagues in length. It is more certain, that, since that time, things have changed still more. The sand thrown up by the sea and winds have formed, between the island of Pharos and the site of ancient Alexandria, a tongue of land two hundred fathoms in breadth, upon which the modern city has been built. It has blocked up the nearest mouth of the Nile, and reduced the lake Mareotis to almost nothing, while, during the same period, the alluvial matter carried down by the Nile, has been deposited along the rest of the shore, and has greatly increased its extent.

The ancients were not ignorant of these changes. Herodotus says, that the Egyptian priests regarded their country as a gift of the Nile. It is only in a manner, he adds, within a short period, that the Delta has appeared. Aristotle observes, that Homer speaks of Thebes as if it had been the only great city in Egypt; and nowhere makes mention of Memphis. The Canopian and Pelusian mouths of the Nile were formerly the principal ones: and the coast extended in a straight line from the one to the other; and in this manner it still appears in the charts of Ptolemy. Since then, the water has been directed into the Bolbitian and Phatnitic mouths; and it is at these entrances into the sea that the greatest depositions have been formed, which have given the coast a semicircular outline. The cities of Rosetta and Damieta, which were built upon these mouths, close to the edge of the sea, less than a thousand years ago, are now two leagues distant from it. According to Demaillet, it would only have required twenty-six years to form a promontory of half a league in extent before Rosetta.

An elevation is produced in the soil of Egypt, at the same time that tills extension of its surface takes place, and the bed of the river rises in the same proportion as the adjacent plains, which makes the inundations of every succeeding century pass far beyond the marks which it had left during the preceding ones. According to Herodotus, a period of nine hundred years was sufficient to establish a difference of level amounting to ten or twelve feet. At Elephantia, the inundation at present exceeds by seven feet the greatest heights which it attained under Septimus Severus, at the commencement of the third century. At Cairo, before it is judged sufficient for the purpose of irrigation, it must exceed, by three feet and a half, the height which was necessary in the ninth century. The ancient monuments of this celebrated land have all their bases more or less buried in the soil, The mud left by the river even covers, to a depth of several feet, the artificial mounds on which the ancient towns were built.

The delta of the Rhone is not less remarkable for its increase. Astrut gives a detailed account of it in his Natural History of Languedoc; and proves, by a careful comparison of the descriptions of Mela, Strabo and Pliny, with the state of the places as they existed at the commencement of the eighteenth century, taking into account the statements of several writers of the middle age, that the arms of the Rhone have increased three leagues in length in the course of eighteen hundred years; that similar additions of land are made to the west of the Rhone; and that a number of places, which were situated, six or eight hundred years ago, at the edge of the sea or of large pools, are now several miles distant from the water.

Any one may observe in Holland and Italy, with what rapidity the Rhine, the Po, and the Arno, since they ha ve been confined within dikes, raise their beds, advance their mouths into the sea, forming long promontories at their sides; and judge, from these facts, how small a number of ages was required by these rivers to deposit the low plains which they now traverse.

Many cities, which were flourishing sea-ports at well known periods of history, are now some leagues inland : and several have even been ruined, in consequence of this change of position. The inhabitants of Venice find it exceedingly difficult to preserve the lagunes, by which that city is separated from the continent; and notwithstanding all their efforts, it will be inevitably joined to the mainland.

We know, from the testimony of Strabo, that Ravenna stood among lagunes in the time of Augustus, as Venice does now; but at present Ravenna is a league distant from the shore. Spina had been built by the Greeks at the edge of the sea; yet in Strabo's time it was ninety stadia from it, and is now destroyed. Adria in Lom- bardy, which gave name to the Adriatic sea, and of which it was, somewhat more than twenty centuries ago, the principal port, is now six leagues distant from it. Fortis has even rendered it probable that, at a more remote period, the Euganian Mountains may have been islands.

M. de Prony, a learned member of the Institute, and inspector-general of bridges and roads, has communicated to me some observations which are of the greatest importance, as explaining those changes that have taken place along the shores of the Adriatic Having been directed by government to investigate the remedies that might be applied to the devastations occasioned by the floods of the Po, he ascertained that this river, since the period when it was shut in by dikes, has so greatly raised the level of its bottom, that the surface of its waters is now higher than the roofs of the houses in Ferrara. At the same time, its alluvial depositons have advanced so rapidly into the sea, that, by comparing old charts with the present state, the shore is found to have gained more than six thousand fathoms since 1604, giving an average of a hundred and fifty or a hundred and eighty, and in some places two hundred feet yearly. The Adige and the Po, are at the present day higher than the whole tract of land that lies between them; and it is only by opening new channels for them in the low grounds, which they have formerly deposited, that theo disasters which they now threaten may be averted.

The same causes have produced the same effects along the branches of the Rhine and the Meuse; and thus the richest districts of Holland have continually the frightful view of their rivers held up by embankments at a height of from twenty to thirty feet above the level of the land.

M. Wiebeking, director of bridges and highways in the kingdom of Bavaria, has written a memoir upon this subject, so important as to be worthy of being properly understood, both by the people and the government, in all countries where these changes take place. In this memoir, he shows that the property of raising the level of their beds is common in a greater or less degree to all rivers.

The additions of land that have been made along the shores of the North Sea, have not been less rapid in their progress than in Italy. They can be easily traced in Friesland and in the country of Groningen, where the epoch of the first dikes, constructed by the Spanish governor Gaspar Robles, is well known to have been in 1570. An hundred years afterwards, land had already been gained, in some places, to the extent of three quarters of a league beyond these dikes and even the city of Groningen, partly built upon the old land, on a limestone which does not belong to the present sea, and in which the same shells arc found as in the coarse limestone of the neighbourhood of Paris, is only six leagues from the sea. Having been upon the spot, I am enabled to adduce my own testimony in confirmation of facts already well known, and which have been so well stated by M. Deluc. The same phenomenon may be as distinctly observed along the coasts of East Friesland, and the countries of Bremen and Holstein, as the period at which the new grounds were inclosed for the first time is known, and the extent that has been gained since can be measured.

This new alluvial land, formed by the rivers and the sea, is of astonishing fertility, and is so much the more valuable, as the ancient soil of these countries being covered with heaths and peat-mosses, is almost everywhere unfit for cultivation. The alluvial lands alone produce subsistence for the many populous cities that have been built along these coasts, since the middle age, and which perhaps would not have attuned their present flourishing condition, without the aid of the rich deposits which the rivers had prepared for them, and which they are continually augmenting.

If the size which Herodotus attributes to the Sea of Asoph, which he makes equal to the Euxine, had been less vaguely indicated, and if we knew precisely what he meant by the Gerrhus, we should there find strong additional proofs of the changes produced by rivers, and the rapidity with which they are made; for the alluvial depositions of rivers alone have, since the time of Herodotus, that is to sa , in the course of two thousand and two or three hundred years, reduced the Sea of Asoph to its present comparatively small size, shut up the course of the Gerrhus, or that branch of the Dnieper which had formerly joined the Hypacyris, and discharged its waters along with that river into the gulf called Carcinites, now the Olu-Degnitz, and reduced the Hypacyris itself to almost nothing. We should possess proofs no less strong of the same kind, could we be certain that the Oxus or Sihoun, which at present discharges itself into the lake Aral, formerly reached the Caspian Sea. But we are in possession of facts sufficiently conclusive on the point in question, without adducing such as are doubtful, and without being exposed to the necessity of making the gnorance of the ancients in geography the basis of our physical propositions.

Progress of the Downs.

The downs or hillocks of sand winch the sea throws up on low coasts, when its bottom is sandy, have already been mentioned. Wherever human industry has not succeeded in fixing these downs, they advance as irresistibly upon the land as the alluvial depositions of the rivers advance into the sea. In their progress inland, they push before them the large pools formed by the rain which falls upon the neighbouring grounds, and whose communication with the sea is intercepted by them. In many places they proceed with a frightful rapidity, overwhelming forests, buildings, and cultivated fields. Those upon the coast of the Bay of Biscay have already overwhelmed a great number of villages mentioned in the records of the middle age; and at this moment, in the single Department of the Landes, they threaten ten with inevitable destruction One of these villages, named Mimisan, has been struggling against them these twenty years, with the melancholy prospect of a sand-hill of more than sixty feet perpendicular height visibly approaching it.

In 1802, the pent up pools overwhelmed five fine farming establishments at the village of St Julian. They have long covered up an ancient Roman road leading from Bourdeaux to Bayonne, and which could still be seen forty years ago, when the waters were low. The Adour, which is known to have formerly passed Old Boucaut, to join the sea at Cape Breton, is now turned to the distance of more than two thousand yards.

The late M. Bremontier, inspector of bridges and highways, who conducted extensive operations upon these downs, estimated their progress at sixty feet yearly, and in some places at seventy-two feet. According to this calculation, it will only require two thousand years to enable them to reach Bourdeaux; and, from their present extent, it must have been somewhat more than four thousand years since they began to be formed.

The overwhelming of the cultivated lands of Egypt, by the sterile lands of Libya, which are thrown upon them by the west wind, its a phenomenon of the same nature with the downs. These sands have destroyed a number of cities and villages, whose rains are still to be seen and this has happened since the conquest of the country by the Mahometans, for the summits of the minarets of some mosques are seen projecting beyond the sand. With a progress so rapid, they would, without doubt, have filled up the narrow parts of the valley, if so many ages had elapsed since they began to be thrown into it; and there would 1no longer remain an thing between the Libyan chain and the Nile. Here, then, we have another natural chronometer, of which it would be as easy as interesting to obtain the measure.

Peat-Mosses and Slips.

The turbaries, or peat-mosses, which have been found so generally m the northern parts of Europe, by the accumulation of the remains of sphagna and other aquatic mosses, also afford a measure of time. They increase in height in proportions which are determinate with regard to each place. They thus envelope the small knolls of the lands on which they are formed; and several of these knolls have been covered over within the memory of man. In other places the peat-mosses descend along the valleys, advancing like glaciers, but differing from them in this respect, that, while the glaciers melt at their lower part, the progress of the peat is impeded by nothing. By sounding their depth down to the solid ground, we may estimate their age; and we find, with regard to these peat-mosses, as with regard to the downs, that they cannot have derived their origin from an indefinitely remote period. The same observation may be made with regard to the slips or fallings, which take place with wonderful rapidity at the foot of all steep reeks, and which are still very far from having covered them. But as no precise measures have hitherto been applied to these two agents, we shall not insist upon them at greater length.

From all that has been said, it may be seen, that nature uniformly speaks the same language, everywhere informing us that the present order of things cannot have commenced at a very remote period. And, what is very remarkable, mankind everywhere speaks as nature, whether we consult the received traditions of the various nations, or examine their moral and political state, and the intellectual attainments which they had made at the period when their authentic records commence.

www.ingramcontent.com/pod-product-compliance
Lightning Source LLC
Chambersburg PA
CBHW072312200526
45168CB00014B/1412